D0115933

The Monkey in the Mirror

Also by Ian Tattersall

Man's Ancestors: An Introduction to Primate and Human Evolution (1970)

Lemur Biology (1975) (ed. with R.W. Sussman)

The Primates of Madagascar (1982)

The Myths of Human Evolution (1982) (with Niles Eldredge)

Encyclopedia of Human Evolution and Prehistory (1988, revised 2000) (ed. with Eric Delson, J. A. Van Couvering, and Alison Brooks)

The Human Odyssey: Four Million Years of Human Evolution (1993)

Lemurs of Madagascar: A Field Guide (1994) (with R. A. Mittermeier, W. R. Konstant, D. M. Meyers, R. B. Mast, and Stephen Nash)

Primates: Lemurs, Monkeys and You (1995)

The Fossil Trail: How We Know What We Think We Know About Human Evolution (1995)

The Last Neanderthal: The Rise, Success and Mysterious Extinction of Our Closest Human Relative (1995, revised 1999)

Becoming Human: Evolution and Human Uniqueness (1998)

Extinct Humans (2000) (with Jeffrey Schwartz)

Ian Tattersall

THE Monkey IN THE Mirror

Essays on
the Science of
What Makes
Us Human

Harcourt, Inc.
New York San Diego London

www.HarcourtBooks.com

Library of Congress Cataloging-in-Publication Data
Tattersall, Ian.
The monkey in the mirror: essays on the science of what makes us human/Ian Tattersall.—1st ed.
p. cm.
ISBN 0-15-100520-6
1. Human evolution. 2. Science. I. Title.
GN281.T365 2002
599.93'8—dc21 2001024122

Text set in Granjon
Designed by Linda Lockowitz

First edition
A C E G I K J H F D B

Printed in the United States of America

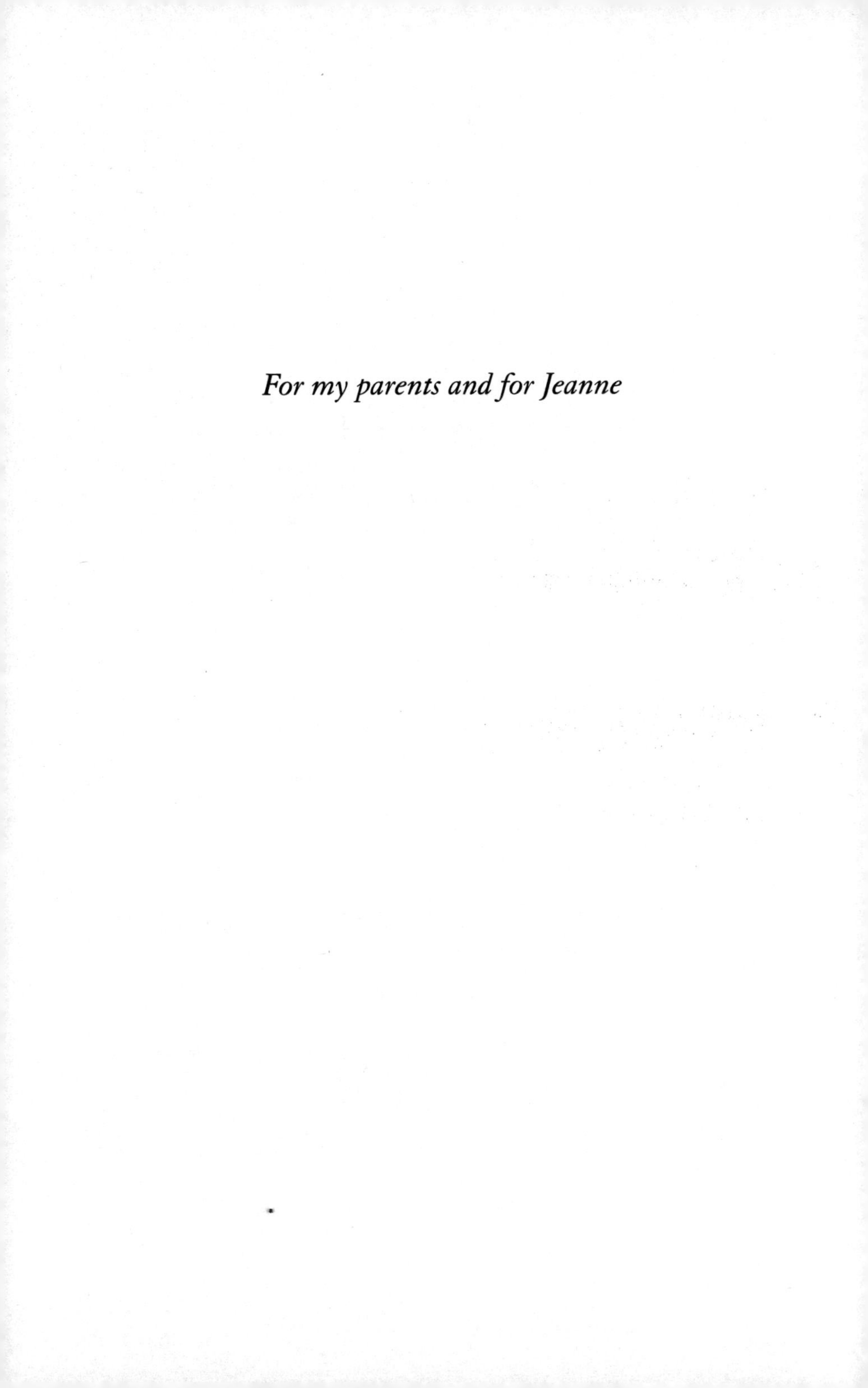

For my parents and for Jeanne

Contents

Preface ix

Chapter One
What's So Special about Science? 1

Chapter Two
Evolution: Why So Misunderstood? 29

Chapter Three
The Monkey in the Mirror 56

Chapter Four
Human Evolution and the Art of Climbing Trees 79

Chapter Five
The Enigmatic Neanderthals 107

Chapter Six
How Did We Achieve Humanity? 138

Chapter Seven
Written in Our Genes? 169

Chapter Eight
Where Now? 185

Preface

It's been an unusual experience for me to write a series of loosely interconnected essays on human evolution and related subjects. When one writes a conventional book, one inevitably follows a predetermined progression, with its own logic and sequence. Essays, on the other hand, don't impose that kind of discipline; they take you where they will, and do not necessarily lead on inexorably from one to the next. This has made the process of writing this book both more interesting and more frustrating than I had expected; and it has led to what for me was an unexpected discovery—although I realize in hindsight it shouldn't have come as a surprise. For, just as all roads used to lead to Rome, it seems that it is impossible to write about anything in human biology without some explicit reference to the process of evolution itself. The great geneticist Theodosius Dobzhansky knew this decades ago, when he wrote that "nothing in biology makes sense except in the light of evolution." But nowadays we tend to be mesmerized by the *results* of four billion years of evolutionary history on Earth, and to take for granted the process by which those results came about. I had thought that one essay on the evolutionary process would be ample to convey the basics of evolution—at least as I see it—to my readers; but instead I found myself coming back to the subject over and over again, throughout the book. Whether the topic was how we came to stand upright; or placing human beings in the context of our closest living relatives; or trying to explain how *Homo sapiens* made that last fateful step to becoming human; or examining whether it is indeed possible to blame our "hunting/gathering" genes for our often bizarre behaviors; or even evaluating the prospects for the human future—I inevitably found myself drawn back to the much-misunderstood process that lies behind all of these things. I hope, then, that readers will forgive me for a small amount of repetition, and for what may look like a mildly obsessive concern with process at the expense of product. The two really are related, and neither can be understood in isolation from the other.

Because much of what I say in this book is at odds with the 1950s-era beliefs about evolution that are still widely accepted, I should also reveal right away exactly where it is that I am coming from. I started my paleoanthropological studies in the early 1960s, a time when it was taken for granted that human evolution had consisted of little more than a long, singleminded trudge from primitiveness to perfection. This prevailing view was supremely linear: Over the eons *Australopithecus* gave rise to *Homo erectus* gave rise to *Homo sapiens*—all under the beneficent eye of natural selection. Quite apart from the fact that this story is not only extremely boring but seriously misrepresents what typically happens in the course of evolution, accepting this schema involved (even then) sweeping under the rug a huge amount of evidence of diversity in the human fossil record. It was fortunate, then, that at an early stage of my career I was sidetracked into the study of Madagascar's lemurs, perhaps the loveliest and most charismatic of all human relatives (they/we are all primates). There is only one kind of human being on Earth today, and from our insular standpoint we seem to feel that it is somehow inevitable, or even appropriate, that we are alone in the world. But among the lemurs it was impossible to ignore diversity. Madagascar is home to fifty different kinds of lemur, and until quite recently to perhaps as many as twenty more. Wherever you look in this unfortunately devastated island you are confronted with evidence for diversity. Nobody studying its attractive denizens could possibly wish to arrange them in a linear order; instead, the obvious question is: How did they diversify so remarkably in their isolated outpost? What's more, the question is essentially the same when you turn from the lemurs to other primate, even mammal, groups. The living world *is* diverse, and wherever you look the predominant signal is one of variety rather than of continuity. Human beings are distinctly unusual in their stately isolation. Well, there's no denying that this does imply there is something special about *Homo sapiens*. Perhaps we are playing the evolutionary game today by a new set of rules. But whether the same thing applied to the predecessors of our species is another question, and I strongly suggest that it did not. We will never properly understand our antecedents, and how we ourselves became as we are, without placing hominids in a normal biological context. By isolating ourselves on a pedestal, we forfeit the opportunity to do this.

The period from the mid-1970s to the mid-1980s was a difficult era—politically, bureaucratically, and economically—in much of the developing world. An adequate account of my many field misadventures during this time, especially in Madagascar and the Comoro Islands, would take another book. The upshot of this period of upheaval, though, was that I was eventually forced to abandon my study of lemurs in the only places where they existed in the wild. I was greatly distressed by this, for I would have been happy to devote my entire career to the contemplation of these wonderful animals. But in retrospect I certainly cannot complain about the outcome, for I was obliged to switch my attention back to hominid evolution. And I returned to this preoccupation with a changed perspective. The lemurs had told me a tale of diversity; and looking at the human fossil record, which had been steadily expanding over this period, taught me the same thing about hominids. There are a lot of different hominids out there in the record, and they certainly don't all fall into a single straight line. In fact, the hominid story is just like that of most other groups of organisms: a tale of continual evolutionary experimentation, with constant origination of new species, triage among those species by competition, and the extinction of the unfortunate—very often without any reference at all to traditional notions of adaptation and natural selection.

If there is one unifying theme in this also rather diverse work, then, it is this: What do we know about diversity in our family, living and extinct, and—very importantly—how do we know about it? From the beginning I have been encouraged to pursue this interest by my editor at Harcourt, Jane Isay, without whose encouragement I would never have presumed to undertake this book, and without whose insights it would not have been as good as it is. I would like to thank Beth Harrison for her sensitive copyediting, and Jennifer Aziz for keeping things on track. Among those many colleagues from whose wisdom and experience I have particularly benefited over the years, Niles Eldredge, Jeff Schwartz, and Bob Sussman occupy a special place. Ken Mowbray, as always, contributed valuable practical help, and my wife, Jeanne Kelly, was endlessly supportive and forbearing. Finally, I would like to take this opportunity to thank all those members of the reading public who read and enjoyed my previous books, and who wrote to tell me of their reactions. Science is of little broader use if it remains the exclusive province of scientists, and it is a great encouragement to individuals like me to know that there are smart people out there who care about the research we do.

THE Monkey IN THE Mirror

Chapter One

What's So Special about Science?

Just south of Tanzania's border with Kenya, the teeming Serengeti Plains are sundered by a giant thirty-mile-long and three-hundred-feet-deep gash. This dramatic erosional feature is Olduvai Gorge, an iconic name in human prehistory. Seen from the air in the clear evening light, and especially on those occasions—not rare—when its rim is touched with gold, the Gorge exudes a soft, magical aura that approaches the mystical. During the height of the day, in contrast, this rugged ravine is a hellhole of rough, heat-reflecting rock, whose variegated colors are almost bleached out by the scorching intensity of the sun's rays. This is not a place where most rational people would choose to tarry any longer than absolutely necessary. Yet, if you carefully approach the Gorge's dizzying edge at certain times of the year, you are almost guaranteed to see, far below, some tiny figures delving into the hot earth at the Gorge's bottom. Chances are that these are not Maasai tribesmen busy driving their cattle from one ephemeral water puddle to another. No, they are most likely paleoanthropologists, scientists involved in uncovering evidence of the human past. For the rocks exposed in the walls of the Gorge are witnesses to almost two million years of geological and human history, and have attracted paleontologists, archaeologists, and others ever since the Gorge's discovery in the early years of the twentieth century. Most famously, this is the place where Mary and Louis Leakey discovered the remarkable *Zinjanthropus* skull in 1959, and from which *Homo habilis*—"handy man"—was reported a couple years later.

Looking at those half-naked, sunburned figures sweating away far below, you can't help but reflect that they hardly conform to the popular stereotype of the scientist. Ask most people for their image of a typical scientist, and chances are they will conjure up visions of a white-coated figure tending high-tech instruments in a spotless air-conditioned laboratory, or covering a well-worn blackboard with elaborate mathematical formulae in an ivy-covered building on some ancient campus. And, at least to a certain extent, this is not inaccurate. Some scientists, possibly even a majority, do look and behave this way, at least some of the time. But scientists are people, and they are quite as diverse as any other category of human being. A lot of scientists, such as those we glimpsed at Olduvai, are principally field-workers, as are many geologists, botanists, archaeologists, ecologists, primatologists, and a whole host of others. Some, though, even practitioners of those very same fields, rarely leave the comfort and safety of the laboratory. What do these disparate individuals have in common? What is it, exactly, that both gives science its unity and sets it apart from the myriad other forms of human endeavor? Well, let's put one common misapprehension to rest right away. It is not that they are all diligently applying "the scientific method." Indeed, there *is* no single scientific method. Scientific methods of course there are, in abundance; and methodologies lie at the heart of the immense variety of different things that scientists do. But different techniques are used for tackling different problems, and there is no particular method that will give you the key to all types of scientific inquiry.

The Nature of Science

Perhaps, then, if it is not a method of investigation that gives science its unity, it might be better to start with what science is *not.* And this is important because, in America today, more perhaps than in any other advanced nation, there is a widespread mistrust of science that is largely born of a lack of understanding of what science actually is and is not. On the one hand, we are happy to benefit from the discoveries of science, and from their applications in the biomedical and technological realms that in the twentieth century have brought so many of us longer lives and vastly greater comforts. On the other hand, though, we tend to fear scientific advance, or at the very least to look upon it askance. At one end of the negative spectrum we imagine practical disaster, as in some of the more extreme scenarios invoked by those who oppose genetically modified crops. At the other, we find the political imposition of the tenets of creationism taught as science in schools, as if mainstream scientific beliefs were somehow intrinsically opposed to those of religion—which they most emphatically are not. Religion is based on faith, while science is grounded in doubt; and although both religion and science do deal with questions of origin, the province of religion is ultimate cause, while the causes investigated by science are proximate ones. It seems highly unlikely that science will ever penetrate the ultimate mysteries—and it probably won't even seriously try, although of course it's not hard to find a scientist willing to pontificate about virtually any subject you want.

The creationist misunderstanding, in particular, stems from the notion that science is an authoritarian system of belief that tells us in absolutist terms how and what the universe is. Here the rather alienating image of the white-coated scientist, rigorously quantifying the world in the remoteness of his laboratory, doesn't help at all. It is, indeed, hard to imagine a more intimidating authority figure than Mr. White-Coat, his clipboard covered with incomprehensible hieroglyphics that the uninitiated—you and I—will never understand. Perhaps it would be better if the more accessible image of the overheated and disheveled field-worker had wider currency, given how much easier it would be for most people to identify with. What's important to realize, however, is that either way the pursuit is essentially the same; for scientific knowledge, whether developed through laboratory experimentation or by outdoor observation of nature, is inherently provisional. And most scientists are only too ready to acknowledge this reality. No scientist who thinks twice about the matter is going to claim that he or she is really in pursuit of "the truth," or has any hope of demonstrating it definitively, although this is admittedly a usage that is all too easy to slip into. All any honest scientist is really trying to do is to *approximate* the truth, in the realization that ultimate truth is unknowable through scientific means, and that the knowledge he or she generates is invariably susceptible to modification.

This is why the familiar mantra of creationists, that "Darwinism is *only* a theory," goes simply to show how deeply these well-meaning people misunderstand what science is all about. For in the most profound of senses, *all* scientific knowledge is "only a theory." Religious belief is a matter of revealed truth and is thus (within interpretive limits) unchanging. Scientific belief, on the other hand, is even in principle only valid as long as it can resist attempts to show that it is wrong. Indeed, while science as a whole embodies a profound feeling that progress *can* be made—for otherwise, what is the point of the whole enterprise?—there is no way in which we can make scientific progress unless we can demonstrate that what we believed yesterday, or even today, is somehow erroneous or, at the very least, incomplete. And scientists tomorrow, of course, will be just as sceptical of today's beliefs.

The way I like to look at it, the core of the scientific endeavor amounts to no more than the corporate effort to describe nature and its workings as accurately as possible. Some of these descriptions are very strongly substantiated and corroborated, and are unlikely to change dramatically with time. The 1953 decipherment of the structure of DNA, the molecule inside the cell which transmits hereditary information between generations and directs the development of every new individual, is a case in point. Here was a truly epoch-making discovery, which gave birth to the entire new field of molecular genetics; and nobody today or in the foreseeable future is going to dispute the basic description of the DNA molecule as double-stranded, helically coiled, and consisting of long strings of four nucleotides that join together in pairs. Yet this was only the beginning. Knowing the basic structure of the molecule was simply a prologue to learning how it carries out its dual function; and while a huge amount has been learned since 1953, a great deal of uncertainty still exists. For example, most of the DNA inside each human cell consists of "junk" whose functions are poorly understood—or which may even have no function at all. The hard graft is done by the genes, which compose only a small fraction of each individual's DNA. But even after many years of intensive research we still have no clear handle on how many genes there are in the human genome, despite the fact that a draft description of the entire human genetic complement has lately been announced. Latest estimates of the number of human genes range from 30,000–35,000 all the way up to 120,000 (with a distinct bias toward the lower end); and if one thing in science is clear it is that we are facing an indefinite future of fine-tuning of these numbers.

This is only one example, but it's an unavoidable fact that the solution of one scientific problem regularly leads to the identification of others that are probably equally tricky if not more so. Throughout the history of science, the successful climbing of an intellectual summit has always revealed new peaks beckoning beyond. In an almost unimaginably complex world, in other words, it seems pretty much inevitable that the resolution of any scientific problem will lead to the identification of another, if not to many more. The upshot is that what scientists are emphatically *not* doing is steadily building up a picture of "the truth." If this were the case, the practice of science would be rather like doing a giant jigsaw puzzle, in which each new piece put in its correct place becomes a permanent part of a gradually emerging but still essentially static picture. By implication, the picture will at some stage be complete, and from day to day what is already established will not change. The fact is, though, that even if there must in principle be a dead-accurate description of the world out there somewhere, like the picture on the puzzle box, science is not equipped to identify it as such. Instead, the process of science is much more like negotiating a hugely complex maze, and the history of science is littered with false starts and backtrackings. New ideas—including new descriptions of nature, mostly tiny corners of it—are proposed, and once those ideas and observations are out there in the public arena, they can be tested. Very often these new notions, usually proposed to correct deficiencies identified in earlier ideas, or to accommodate new data that don't fit with whatever was previously believed, are—or will turn out to be—wrong, either in detail or in their entirety. The neat part, though, is that this doesn't matter at all. In science it is no crime to be wrong, unless you are (inappropriately) laying claim to the truth. What matters is that science as a whole is a self-correcting mechanism in which both new and old notions are constantly under scrutiny. In other words, the edifice of scientific knowledge consists simply of a body of observations and ideas that have (so far) proven resistant to attack, and that are thus accepted as working hypotheses about nature. This may sound like a rather rough-and-ready way of proceeding; but clearly, in the mere two or three centuries that have elapsed since science as we recognize it today began to come into existence, it has brought us a remarkably long way. For evidence of this we need look no farther than the totally unprecedented efficiency with which we are now able to manipulate the world around us. Few would dispute that science truly has revolutionized all of our lives, in a way in which no other approach to knowledge has ever managed to do.

Falsifiability

For this system of provisional knowledge to work, however, it's necessary that, to the extent possible, scientific hypotheses be proposed in such a way that they are at least potentially falsifiable—provable to be wrong. Nonscientific statements about the world can simply be judged by the criterion of plausibility, which is fine in its place; but a scientific statement has to be subject to disproof if it's wrong or lacking something, as the implicit assumption will always be that in some way it is. In other words, for a notion to be accepted as fully scientific, it has to be one that, if wrong, can be shown to be wrong by more than simple assertion. Scientists (possibly it would be more accurate to say "most scientists," humans being what they are) are not out to *prove* anything, however much notions of "scientific proof" may be bandied about in some circles. They are simply trying to come up with descriptions that resist *dis*proof in terms of what is already known or believed. Of course, many areas of science depend heavily on techniques of mathematical description, and mathematicians are frequently in search of "proofs" of one conjecture or another. But it is a mistake to confuse quantifiability with objectivity. The various branches of mathematics are essentially systems of logic that are based on axiomatic starting assumptions. And while scientists find the techniques of mathematical description very helpful in characterizing the world, they themselves cannot start from assumptions. They have to start from what they know about the world, in the knowledge that what they think they know is always subject to change.

That's the theory, anyway, and it applies pretty well in the experimental sciences such as physics. There scientists generally start either from new hypotheses that they hope describe the world accurately, or from established notions that seem to be becoming a little wobbly. These they then test against new data garnered by experiment and observation, often expressly for this purpose. There is, however, another category of sciences that few reasonable observers will deny is "scientific," but in which the nature of the phenomena being studied precludes those involved from taking the experimental route. These are the "historical sciences," most notably evolutionary biology. We can study the *functioning* of the hereditary molecules within each cell by conventional scientific methods of experimentation, but what we cannot test directly by setting up experiments is the *history* of those molecules: exactly how they came to be as they are, and how their properties came to be distributed in nature in the way that we observe. These are matters of history on an immensely long time scale, and those histories can never be replicated in the laboratory. Fortunately, though, there is a way around this. Experimental scientists make predictions about the outcomes of their experiments, and then compare the data gathered against those predictions. And so do evolutionary biologists. The difference is merely that in their case the experiments have already been made, long ago.

The central prediction that emerges from evolutionary theory is based on the common descent of all life forms. If all life is descended from a single common ancestor, then we should expect to see a "nested" pattern of resemblance among the varied descendants of that ancestor. It should be possible to represent all of life in a single branching diagram that ramifies upwards from a single ancestor at the bottom. Actually, things seem to have been a bit more complicated than this, at least at the beginning of the history of life. All life may not, in fact, have had a singular origin (and why should it; if simple self-replicating molecules could emerge once, then why not multiple times in a largely competition-free age?), whilst, early on, at least one significant new form of life may have risen from a combination of old ones. The important thing, though, is that we would never have realized or have begun to understand this if we had not started from a hypothesis—that all life did have a common origin—which we could test and refine by reference to the structure of the living world around us. And in any event, subsequent to the establishment of the major groups of living organisms, we do find a very strong overall signal when we compare the distribution of characteristics among the presumed descendant forms. People have realized this since time immemorial, of course; "folk taxonomies" have long reflected the realization that the living world is organized into groups-within-groups that can be defined on the basis of characteristics that their members share; and nobody who has any familiarity with the living world has any problem distinguishing a bird from a bat or a flying fish. What science allows us to do is to move beyond such levels of generality and more precisely to specify the relationships among organisms—which sometimes turn out unexpectedly; who would until fairly recently have claimed that lungfish are more closely related to cows than they are to salmon?

Among the very few philosophers of science who have been taken at all seriously by scientists themselves was the late Sir Karl Popper. Popper was the leading proponent of the notion of falsifiability as the crux of scientific ideas, and it was principally for this that he became famous. Especially in his early days, however, Popper took a rather dim view of evolutionary biology as science, declaring that there are "no evolutionary laws." What he meant by this was actually that none of the "evolutionary laws" he chose as examples was in fact in the form of a scientific law, in the sense that such statements must be both invariable and true everywhere. Fair enough. What's more, he regarded such notions as "all earthbound life evolved from a single progenitor" as probability statements, rather than as general laws—in which, as we've seen, he was most likely correct as well. But here Popper was missing his own point. If scientific knowledge is provisional, as the falsifiability criterion implies, then we probably shouldn't be looking for "laws" at all, however tempting such a pursit might be. "Rules of thumb" might be a better term for most scientific generalizations. Later in his career, Popper softened his stance somewhat, describing evolutionary biology as a "metaphysical research program." Whatever this actually means, it falls upon the ear as vaguely derogatory, and Popper's description was seized upon eagerly by the foes of evolution. Metaphysics (whatever it may actually be) certainly sounds like the antithesis of science. In Popper's universe, however, this change of terminology amounted to something of a compliment, since for him it carried the implication that Darwinism provided "a possible framework for testable scientific hypotheses." What's more, Popper finally came around to the view that "the survival of the fittest," which he had initially attacked as a tautology, was in principle testable, if with practical difficulties. Here he was probably right first time, and of this more later. But in any event, the unfolding of an evolutionary history is the best explanation we have—and the *only* predictive one—for the pattern of life we see around us. Indeed, the only possible alternative is that of special creation, which yields no predictions. If the living world was created by a supernatural being, then the world is the way it is simply because that being wanted it this way. Fine, if this is what you want to believe; but just don't try to dress it up as science. The notion of evolution predicts the nested pattern of relationships we find in the living world; supernatural creation, on the other hand, predicts nothing. It is concepts of this latter kind that are truly untestable. And what else is faith about, after all?

Of course, this notion of falsifiability is inherently incomplete. It deals with how ideas should be treated once they are out there in the scientific arena, and how they should be posed so that they can be evaluated. But it also begs an obvious question: Where do the ideas, good or bad, come from in the first place? Well, there are no rules for human creativity—how could there be?—and science depends on creative thought and intuition quite as much as any other branch of human endeavor. When we consider the origin of truly new ideas in science we are, essentially, dealing with the mysteries of human cognition. "Eureka!" is a reaction that is familiar in science (though maybe not as familiar as many of us would like!); but it is, alas, not something that can be consciously conjured up.

Science as a Collective Enterprise

So far, I have been speaking mostly of the kind of science practiced by individual scientists, or by teams of them. But we should never forget that science is above all a huge, worldwide, collective enterprise. While it is possible with some effort to imagine a world without scientists, and more easily a world with many millions of them, short of a post–nuclear holocaust scenario it's literally impossible to imagine a world with just one scientist. From well before 1660, when the savants who founded London's Royal Society gathered regularly to compare their observations of nature, science has been recognized as a corporate endeavor. Sir Isaac Newton was not the first to have said, three centuries ago, that if he had seen farther it was because he had stood on the shoulders of giants. But his classic remark encapsulates a basic verity of the scientific process: All science has to start from what is currently known about the world, or from what is believed about it. This in turn helps to explain why it is so difficult to put one's finger on the origins of science itself. What is abundantly clear, however, is that what is accepted as scientific knowledge changes and expands, and as science has enlarged it has become compartmentalized into a huge variety of fields. Science is thus an enormous body of knowledge and belief that has been accumulated over the centuries, thanks to the efforts of many thousands, even millions, of investigators. But given the enormous inertia of large bodies of anything, how does scientific change occur at all?

The most persuasive and comprehensive account of this process was published some forty years ago by

Thomas Kuhn, the only philosopher of science whose fame rivals Popper's. Kuhn, who started his career as a physicist, was acutely aware of the role of the scientific community as a whole in spurring scientific advance. In his book *The Structure of Scientific Revolutions* Kuhn pointed out that at any one time belief in any particular area of science tends to be dominated by what he called a "paradigm," a generally accepted explanatory framework into which new observations are incorporated as a matter of course. Such paradigms initially become dominant as large numbers of scientists are attracted to them, and away from competing forms of explanation. And they stand at the origin of new traditions of scientific research, as the new paradigm reveals new questions to be explored. As time passes, however, paradigms tend to become ossified into forms of received wisdom, even as new observations about the world accumulate. The tendency among scientists will be to try to understand those new observations within the context of the accepted paradigm; but at some point so many anomalies at variance with received wisdom will have been identified that it will become clear that a new explanatory framework is necessary. It is at such moments that science is ready to witness a "paradigm shift," in which the old framework is rejected in favor of a new one that can more convincingly accommodate the new observations. It is, in other words, the accumulation of anomalies, usually recognized in the course of research programs originally designed to refine accepted beliefs, that forces science to change its paradigms.

This need not be an overnight process, and indeed it rarely is. I was fortunate enough to be studying geology in graduate school right at the moment when the new notion of plate tectonics was being born. At the beginning of the 1960s, it was still generally believed that the basic form of the Earth's surface was essentially static. Of course, evidence for huge numbers of earth movements was visible in the geological record, and phenomena such as the gigantic Krakatoa explosion of 1883, or the 1906 San Francisco earthquake, were only too fresh in memory. However, geological instabilities of this kind were always viewed as essentially local events, even if it was evident that they had taken place on a huge scale; and geologists sought local causes for them, often with enormous sophistication and ingenuity. Around the early 1960s, however, a new generation of geologists began to make observations that tied earthquake zones, volcanism, rifting of continents and seafloors, mountain building, and a host of other geological phenomena to a picture of an Earth's surface that was constantly undergoing change. It turned out that geography is not fixed after all (as some had been hinting from early in the twentieth century, based on the complementarity of the outlines of the Atlantic continents). Instead, the continents represent unstable blocks of relatively light rock that float around on the heavier molten rocks below them. Oceans are formed by the rifting apart of continental blocks, while the huge forces unleashed by collisions between the drifting fragments are responsible for earthquakes and mountain building, and volcanoes result from the escape of molten material from below. If it were not for this constant and often violent process of renewal, the continental surfaces would long ago have subsided and been eroded below the surface of the oceans, and there would be no room for terrestrial life. In a remarkably short time, in other words, a new explanatory framework was developed that for the first time gave us a comprehensive mechanism that knit together a whole variety of apparently diverse geological phenomena.

You might have thought that geologists would have been delighted to see their area of science intellectually unified in this way. Not so. There was tremendous resistance to the new ideas, not just among the old guard, but also among younger colleagues who remained under their influence. This is not surprising, of course, or even reprehensible: It takes a while to detect which way the wind is blowing, and it is difficult indeed to reject principles to which one has devoted one's career. And of course the new field of plate tectonics certainly didn't summarily invalidate the vast bulk of the detailed local observations on which other geological explanations had been founded. What's more, there is considerable inertia built into the process of scientific education. Textbooks take years to update with new knowledge, and it is remarkable how early in the educational process the mindsets of young scientists become established. What one is first taught on a subject tends to have much more influence than whatever one hears subsequently about the same topic. It is thus a huge responsibility for any teacher to present a view of a scientific field to students who are hearing it for the first time. If the teacher is at all effective, he or she will almost certainly inculcate a view of the world that down the line will prove very difficult to modify in the students' minds, however much the evidence changes. Add to this that the role of doubt in the scientific process is far too rarely taught to those who aspire to become practicing scientists, and the potentially oppressive power of received wisdom becomes painfully apparent.

Still, Kuhn was undoubtedly right: Paradigms must change sooner or later, as knowledge accumulates. The piling-up of anomalous observations that cannot be explained by old hypotheses cannot forever be ignored, and must eventually lead to the demise of inadequate explanatory frameworks, however tenaciously these latter may tend to linger. The history of science has borne out this pattern over and over again, and in fact it is not necessarily only new observations that lead to paradigm shift. Some paradigms are essentially intellectual: They are views of how science should be done, and are not dependent on any specific set of observations. Indeed, in my own science of paleoanthropology we are at this very moment in the middle of a paradigm shift of this kind.

Let me explain. When I was in graduate school, my office was a desk in a basement storeroom of a natural history museum. I would watch enviously as visiting scientists pored for hours over fossils that they pulled from drawers in the cabinets that lined the room, making reams of notes and measurements. It seemed that these people—these initiates—knew exactly what they were doing, while nobody had yet taken the trouble to explain to me how to go about studying fossils. I had attended innumerable courses in vertebrate paleontology, of course, but the emphasis was always on the instructor's interpretations of particular fossils, and not on how those interpretations were arrived at. Eventually I plucked up the courage to ask a distinguished paleoanthropologist what the secret of studying fossils was. The answer? "You look at them long enough, and they speak to you." Well, yes, okay. It's true that sheer familiarity with fossils will reveal things about them that nothing else can. But as my colleague Milford Wolpoff once said, "I've spent a lot of time alone with fossils, and none of them ever said a word." And my hearing is no better than Milford's. Of course, it's useless to deny that a largely seat-of-the-pants approach to studying fossils had brought paleontologists a very long way since the early nineteeth century. An intuitive appreciation of anatomical similarities and how they are distributed among living organisms had permitted some very smart people to arrive at a remarkably detailed and accurate description of the diversity of life. But even as I was receiving my rather dusty answer to my innocent question, I felt that surely there must be something more than this in the study of mute fossils.

And, of course, there is. It was great good fortune that took me at the beginning of the 1970s to the American Museum of Natural History, where a revolution in systematics (the science of analyzing relationships among organisms) was getting underway. This was the introduction into American systematics of "cladistic" methods (from the Greek word *clados,* meaning "branch"). Traditionally, the paleoanthropological notion of theoretical rigor was to look at "total morphological pattern" rather than at single characters in deciphering evolutionary relationships. The problem was, of course, that nobody could agree on what total morphological patterns actually were, so competing notions of relationship were impossible to test. Cladistics, in contrast, focuses on individual characters, and recognizes the significance of the distinction between "primitive" and "derived" character states. While common possession of primitive character states (those present in the ancestor of the group) indicates overall group membership, relationships within the group are only specified by the common possession of derived character states. These relationships are represented in branching diagrams called "cladograms." I will spare you the details of how primitive versus derived characters are recognized, and how cladograms are constructed; suffice it to say that, with the advent of cladistics, systematics had finally acquired a truly scientific basis. "A and B possess derived characters not shared with C," which is what the simplest cladogram says, is a truly testable statement, as is the inference that A and B are most closely related by common ancestry. If we go beyond this to more complex (and admittedly more interesting) statements, for example to hypotheses of ancestry and descent, the picture becomes murkier because such notions cannot be corroborated with any certainty; we are back to probability judgments, and unquantifiable ones at that, although we can now see where they are coming from. The narrow-minded might conclude from this that science should stop right there, that when we move away from the strictly testable we are moving beyond the boundaries of science. But of course, there is no reason whatever why scientists should not investigate the murkier and more intractable (not to mention more interesting) areas of human experience—and every reason why they should!

Science and Paleoanthropology

You might imagine that paleoanthropologists would have welcomed the advent of cladistics with open arms. Here, at last, was a method allowing direct comparison between competing ideas about evolutionary relationships. Well, over the years there has been some grudging (and in a few places, even enthusiastic) acceptance, and today cladograms are quite frequently seen in paleoanthropological journals. However, it is hard to avoid the impression that the forms and terminologies of cladistics have been more warmly embraced by many paleoanthropologists than the philosophy and the implicit view of the structure of nature that lies behind this approach. And some colleagues continue to reject cladistics outright, or to insist on misunderstanding it. There is certainly plenty of evidence that a paradigm change in methods of human phylogeny reconstruction has begun, but it will be a long time before cladistics truly becomes received wisdom. And that, as I've hinted, is par for the course—

and is not necessarily a bad thing. Science thrives on diversity, and the more new ideas that are proposed, and are made available to be tested, the better.

Which, of course, is precisely what was going on at Olduvai, where we began this discussion. Some of those tiny figures down there on the Gorge's hot floor might have been paleontologists, looking for direct bony evidence of early humans and the animals they lived among. Others were probably archaeologists, looking for traces of early human activities (and indeed, it was at Olduvai that truly ancient stone tools were identified for the first time). Some may have been geologists, refining their ideas about what the rocks exposed in the Gorge's walls are telling us about past conditions. And yet others might have been taphonomists, scientists who try to elucidate what happens to animals after they die, and thus to understand exactly how the fragmentary evidence of the past has come down to us today. It is very important to know this, for not everything we observe can be taken at face value. At one time, for example, it was believed that a 1.8-million-year-old circle of stones that had been noticed at a site on the Gorge's floor represented the remains of a deliberately constructed windbreak. If so, this was the earliest structure known, antedating anything comparable by over a million years. Closer examination—testing of this hypothesis—showed, however, that the "stone circle" was much more probably the result of shattering and scattering of stones by the roots of a growing tree. In retrospect, this reinterpretation was an early shot in a battle to revise our overall view of the nature of early hominids. These beings had tended to be viewed as little more than primitive, unsophisticated modern humans; and in a limited sense this perspective allowed scientists to use *Homo sapiens* and our own behaviors as a "model" for understanding the behaviors of our forebears. Today, however, it is widely recognized that looking at our predecessors as no more than junior-league versions of ourselves may be profoundly misleading as a guide to understanding the kinds of creatures they were. This is another example, if on a very small scale, of paradigm change, and hence the very simple answer to the question with which we started this essay: What do all those very diverse folks down there in the Gorge have in common? Whatever their precise interests and techniques, they are all engaged in the attempt to expand, refine, and above all to test what we *think* we know about the past—an exercise that will continue in this and other fields of investigation as long as there are scientists.

One other thing, alas, will almost certainly continue in science. Thomas Kuhn had his detractors, but he was undisputably right in his central perception that science is a collective enterprise that has autonomy from all other areas of human endeavor. Put this together with the fact that all scientists are human beings, and you have a recipe for internecine warfare. Even white-coated scientists are not soulless automatons. They have feelings that are quite as strong as other people's; and even though they usually strive toward objectivity they find it as difficult as anyone else to put those feelings aside in their interactions with colleagues. Scientists, in other words, are quite as opinionated and even quarrelsome as practitioners of any other branch of human endeavor. Scientific feuds are not at all uncommon, and this is not only because ideas, however abstract (and fossils, for paleoanthropologists), are an intensely personal possession ("Attack my ideas, and you are attacking me"; "How can *you* presume to tell me how to interpret the fossil *I* found?"). After all, individuals, whatever they do for a living, are always going to find ways to disagree. But discord also arises because science is such a corporate affair, and because to be properly carried out it requires the establishment of institutions. This has two consequences, both in a broad sense ecological, and hence, in the human way of doing things, political. On the one hand, in a world of limited resources there is inevitably a struggle amongst institutions; and on the other, there is an equally inevitable competition for resources within those institutions. This is simply the way of the world, and when we come to look at the evolutionary process we will see that the rest of nature follows a similar pattern. What makes things different among humans is that as individual beings we are so complex, for which read "difficult." And given our human proclivities, it should hardly be surprising that scientists are no different from people in other professions in exhibiting spite, jealousy, and vengefulness as well as cooperation and generosity. Almost any scientist will bend your ear at length with stories of "unprofessional" behaviors exhibited by a colleague. Ironically, though, it is the corporate nature of science that comes to our rescue here. Individuals can and do make huge and essential contributions to science, in terms of innovation and new ideas. Yet the proof of the pudding is in the eating; and ultimately it is the need to carry their colleagues along with them that keeps individual scientists on track. The idea of community really is at the heart of the scientific enterprise—however difficult individual scientists may find it to get along!

Chapter Two

Evolution: Why So Misunderstood?

Evolution. A word that some would have us believe should strike terror, or at least blind reaction, into the heart of every true-blooded American concerned about preserving our treasured way of life. How come? The notion of evolution is, after all, a pretty simple one, and not one that self-evidently impinges markedly on other systems of belief. But there are people out there who are truly convinced that evolution—in essence, the notion that all life forms are broadly descended from a single common ancestor—threatens the very basis of social morality and cohesion. Ironically, the devoutly Christian and socially conservative Charles Darwin would have been appalled by this idea. So what is happening here? The problem seems to lie, as I hinted earlier, with a mistaken notion of what science is. Science, the opponents of evolution maintain, is an authoritarian system of belief, producing axioms that are unchanging for the ages. And of course, all authoritarian systems that promote unchanging views of the world automatically find themselves at odds with other authoritarian systems, especially when those systems are based on a Revealed Word—as numerous religious wars attest. Hence the opposition to ideas of evolution by some (actually a minority) religious believers. But once we realize that science is a system of provisional knowledge, the perspective changes dramatically. Science does not seek to understand ultimate causation, but simply to improve our perceptions of what nature is. Scientists (many of whom are religious believers themselves) are happy, in the main, to leave the really tough questions, such as those surrounding the meaning of human life, to theologians and philosophers, or simply to individual canons of belief. There is, after all, plenty for them to do within the strict rubric of science—and nowhere more so than in the tricky business of figuring out exactly how the extraordinary process of evolution actually works.

Everyone has an evolutionary history, and thus a kind of proprietary interest in the means by which their predecessors emerged from the primordial ooze. But people's notions of this process vary widely, and there is a great deal of popular mythology about it. As I've just suggested, though, the basic notion of evolution itself is pretty straightforward. As elegantly defined by Charles Darwin back in 1859, it is simply "descent with modification." Interestingly, despite the legendary stories of fainting bishops' wives and the immediate scandalization of Victorian society as Darwin's book *On the Origin of Species by Means of Natural Selection* became a bestseller, the general public quite rapidly came to accept the notion that we are related by descent, not just to apes and monkeys but to lizards and lobsters as well. The mute eloquence of the pattern of similarities among the multitude of species that make up the living world was and remains so powerful that, well before the end of the nineteenth century, this fundamental evolutionary idea had become as commonly accepted as it is today (which is to say widely, if not universally). Beyond this point, however, the picture got murkier, for notions of what lay behind this pattern of common ancestry among living forms varied enormously.

Natural Selection

Darwin and his younger colleague Alfred Russel Wallace favored what they termed "natural selection" as the driving force of evolutionary change. Natural selection is a deliciously simple idea, and at first glance it seems almost self-evident (which is why Darwin's friend and defender Thomas Henry Huxley, on hearing of it, is reputed to have said, "How very stupid of me not to have thought of that!"). It is based on the observation that within populations individuals vary among themselves in their physical characteristics, and that those traits in which they vary are often—indeed, usually—inherited from their parents, and will be similarly passed on to their offspring. Some of these heritable variants will be more conducive than others to their possessors' reproductive success, and such favorable attributes will inevitably become commoner in the population as this pattern of differential reproduction repeats itself over many generations. As changes accumulate in this way, the appearance of the populations involved will gradually alter; and over the eons the additive effects of incremental modifications from one generation to the next will result not just in small physical changes but in the emergence of new species, genera, families, and so forth. In this view, time and change are virtually synonymous: The more time, the more change.

The intermediary between natural selection and population change is the notion of "adaptation," quite simply the idea that natural selection favors those individuals whose characteristics best promote life and reproduction in the local environment—whether that environment is stable or changing. Of course, it can be objected that, as commonly used, adaptation is often something of a tautology: Those individuals who survive and reproduce most successfully are by definition those that are best adapted. And there is something in this, as we will see in a moment. But it should be borne in mind that most features will linger in a population as long as they simply don't get in the way, which seems to me to somewhat exceed any useful definition of adaptation. Nevertheless, adaptation is clearly a real phenomenon in evolution, as is demonstrated, for example, by the frequency of "convergence," whereby some remotely related species have developed amazingly similar physical characteristics, or of "mimicry," where, for example, edible species display the distinctive features of inedible ones, to discourage predators.

But in the immediate post–Darwin/Wallace era not everyone was happy to blame evolutionary modification upon natural selection. In particular, as the twentieth century dawned the science of genetics—which is to say, heredity—was in the process of being invented. These were exciting times in many areas of biology, not least in genetics, and at such moments of intellectual ferment there is typically a proliferation of new ideas, as novel possibilities open up and are explored. Naturally enough, then, the newly minted geneticists offered a host of alternatives to classical natural selection. "Mutation pressure," the rate at which new genetic variants were produced in populations, was a popular option as a driving force for evolutionary change. So was "saltation"—a notion of spontaneous major change born of the fact that geneticists occasionally noted the sudden appearance of "sports" (highly distinctive individuals)—among their subjects of study. Some biologists even felt that evolutionary changes were the expression of an innate drive toward a fixed goal, or that environmental shifts somehow induced heritable innovations. Whatever the individual researcher's exact outlook, however, natural selection rarely figured as the major factor in the evolutionary mix. Indeed, it was not until the 1920s and 1930s, with the emergence of the "Evolutionary Synthesis," that natural selection once more began to loom large in formulations of the evolutionary process.

The Evolutionary Synthesis

The Synthesis resulted from the efforts of many people to reconcile evolution with the growing knowledge of genetics, including some inspired mathematical geneticists in both Britain and the United States. But its most influential documents were a series of books produced between 1937 and 1944 by the Columbia University geneticist Theodosius Dobzhansky; and two curators at the American Museum of Natural History, the ornithologist Ernst Mayr and the paleontologist George Gaylord Simpson. Between them, these three scientists spanned those fields of biology most centrally concerned with evolutionary theory (for in those early days, it had yet to become fashionable to dress up inquiries in other areas of biology with the evolutionary veneer that seems almost compulsory today). Remarkably, they contrived to unite their diverse areas of study under a single and attractively simple model of the evolutionary process. Each of these scientists had his own perspective, of course (Dobzhansky was most interested in the genes, Mayr in species, and Simpson in the origin of major groups), and each studied a rather different set of natural phenomena; but when push came to shove all agreed that virtually every major aspect of evolutionary change could be reduced in one way or another to the generation-by-generation modification of gene frequencies in populations, under the influence of natural selection. This is, of course, a model of continuity; and each of these scholars, perceptive naturalists all, was acutely aware of the fact that nature is actually riddled with *dis*continuities. However, they contrived to explain away such awkward observations as special cases of gradual change; and as time passed, the complexities these founding fathers had worried about became gradually lost to sight, as the Synthesis "hardened" into dogma. It was, by the way, in its purest and most dogmatic form that the Synthesis was received, rather late, into the science of paleoanthropology.

One of the most positive effects of the Synthesis—and a hugely valuable one—was to sweep away an enormous entrenched body of scientific mythology. The pre-Synthesis period has aptly been described by Ernst Mayr as "chaotic," with almost as many theories of evolution as there were practitioners of what we would now call evolutionary biology. With the advent of the Synthesis much of the wilder speculation was seen off the scientific stage, to be replaced by a straightforward and coherent body of belief. This elegantly simplified model of evolution would eventually serve as an effective platform for the recognition and incorporation of the complexities that had conveniently been brushed aside in its formulation. Meanwhile, however, with its message of continuity, the Synthesis encouraged the adoption of linear notions of evolutionary history. This was particularly the case in paleoanthropology, where prewar notions of a "bushy" human family tree, with almost every new fossil occupying its own terminal twig, were replaced by that of a single basic lineage leading insensibly from the small-brained early bipeds, through the medium-brained species *Homo erectus,* to the large-brained species *Homo sapiens.*

Species—the basic "kinds" of organisms—are the

fundamental units of evolutionary analysis. And although they present a host of practical problems of recognition, it is widely agreed that species are best defined as the largest groups within which, reproductively, all members are fully compatible. In principle, then, species are clearly circumscribed units. However, according to the dictates of the Synthesis, no matter how readily recognizable they may be in the living biota, species cannot be regarded as "real" entities. For however distinctive they may be in space, they are expected to lose their individuality in time. This is because the Synthesis sees species as little more than segments of lineages that are steadily transforming under natural selection. They should, then, merge insensibly one into the next, with no sharp breaks. Indeed, from time to time during the 1950s and 1960s, one paleontologist or another would express relief that the record was incomplete, since the vagaries of geology produced arbitrary break-points at which species boundaries could conveniently be recognized!

Of course, if the transformational notion were accurate, histories of continuity should clearly show up in the structure of the paleontological record. Yet, if the truth be told, the fossils themselves had never really borne out this expectation. Indeed, Darwin himself had been well aware that the record was rife with discontinuities. He had, however, explained away this awkward fact with the now-familiar claim that the expected intermediates had simply not yet been discovered. In Darwin's day, with a much more limited record than we have now, this was at least a tenable proposition. But well over a hundred years and many millions of fossils later, the essential picture has not changed at all. The more we learn of the fossil record, the sharper the image becomes of species as real, bounded units, with births, histories, and deaths. They tend to appear quite suddenly in the record, to persist for varying but often remarkably long periods of time, and then to disappear as abruptly as they had shown up. Often they are replaced by close relatives, sometimes by unrelated forms. Or they may not be "replaced" at all in the sense that close ecological equivalents take over from them. What we don't often see, however, is compelling evidence of the gradual transition of one species into another.

Punctuated Equilibria

Nonetheless, such was the beguiling elegance of the Synthesis that this didn't seem to matter much. In good Kuhnian fashion, paleontologists were happy to continue cramming their fossils into the kind of linear structure predicted by the Synthesis, and not until the early 1970s was a significant theoretical alternative proposed for their consideration. This was the notion of punctuated equilibria, most famously put forward in a 1972 paper by the American Museum of Natural History curator Niles Eldredge and his fellow invertebrate paleontologist Stephen Jay Gould. While accepting many of the principles on which the Synthesis was founded—and it's important to realize that punctuated equilibria should be viewed as an extension of the Synthesis, rather than as a replacement of it—Eldredge and Gould pointed out that stasis (nonchange) was by far the strongest signal contained in the fossil record. What's more, they were unwilling to view this observation as an artifact of incompleteness. Perhaps, they suggested, the famous gaps in the record were telling us something significant after all. And if so, the message involved was clearly not one of continuity. Instead of being in most cases a gradual, incremental process, evolution is actually a matter of sporadic innovation, as new species suddenly show up in the rocks.

Naturally, this novel formulation of the evolutionary process raised an entirely new set of expectations about the patterns we should anticipate finding in the fossil record as new specimens come to light. And, like all contradictions of received wisdom, the notion of punctuated equilibria predictably had a bumpy ride at first, although it is now very widely accepted, especially outside paleoanthropology. But just as important as the matter of pattern is the emphasis placed by this new view on the significance of speciation—the process by which new species emerge—in the overall evolutionary mix. Previously, speciation had often been explained away as a special case of gradual change; but now it became necessary to confront it as a phenomenon in its own right. Ironically, perhaps, the major work on speciation in animals had been done before the mid–twentieth century by no less than Ernst Mayr, one of the chief architects of the Synthesis. Mayr championed the notion of "allopatric" speciation, whereby new species form when portions of preexisting species become isolated from the rest by ecological accident (seaways form, mountain ranges rise, rivers change course, and so forth). Such events disrupt physical continuity, so that genetic innovations can accumulate independently in the parent and daughter populations. Eventually, the theory goes, these accumulating differences will result in reproductive incompatibility. What's more, since large populations are much more stable genetically than small ones, it is probably in the peripheral isolate rather than in the bigger parental population that the key differences will emerge.

Speciation

Unlike gradual change, for which in essence we only have Darwin's word, we *know* that speciation must occur, since otherwise life could never have diversified as it so clearly has—and diversity is a major theme of the evolutionary record. But in many ways speciation remains the black box of biology, a phenomenon that is of critical importance but that remains poorly understood. What is clear, though, is that what we see as speciation is actually an outcome, rather than a mechanism per se. Speciation involves the appearance of reproductive barriers—"isolating mechanisms"—where there were none before; and many different causes have been identified that can lead to this result. Isolating mechanisms can, for example, lie at the level of the individual gene, or at the level of the chromosomes into which the genes are organized, or even at the level of behavior. Some isolating mechanisms may be "premating," interfering with individuals' ability—or desire—to mate with each other in the first place. Others may be "postmating," making the initial formation of a new individual impossible, or inhibiting the proper subsequent development of the offspring. In some cases the reduced viability of hybrid offspring as juveniles or adults may ensure that they are weeded out of the reproductive population. Whatever the underlying reason, what is important is that the result is the same: the emergence of a new species, or of two species where there was only one before.

It is also crucial to realize that the mechanisms behind speciation are not necessarily the same as those that underlie morphological change within species. Let me explain this a little further. Traditionally, speciation has been thought of as little more than a passive consequence of evolutionary change. Accumulate enough change, and speciation will inevitably result. But if you look around at the way variety in the living world is organized, you realize that this cannot be the case. Some species contrive to diversify enormously without speciating. Perhaps the most dramatic example of this is the domestic dog. There is a huge variety of dogs, both in shape and size; yet all are reproductively compatible, if not directly (think Chihuahuas and Newfoundlands) then via intermediates. Potential gene flow among dogs is uninterrupted, and all remain members of the same species, *Canis familiaris*. Now, you may well object that the enormous variety of man's best friends is a product of artificial breeding. Yes, it certainly is; but although the vast diversification we see within this species is unarguably the result of human intervention, that doesn't change the basic genetic mechanisms that have allowed it. When breeding dogs, humans have simply contrived to stretch these mechanisms to the limit. It is thus evident that at least potentially a vast amount of anatomical variety can be accommodated within a single breeding group; and, with or without human help, it is routine for the great majority of geographically widespread animal species to develop distinctive local variants. It is, indeed, this tendency of populations to diversify that is quite likely the primary motor of anatomical innovation in animal evolution. It's also worth noting that, for all the intensive breeding experiments that geneticists have done on hundreds of generations of such fast-breeding organisms as fruit flies, none has ever contrived to produce a new species, although some pretty odd-looking flies have resulted!

But it is equally true that some closely related species may be reproductively isolated from one another, yet remain hardly distinguishable physically. There are, for instance, species of Amazonian squirrel monkey that can barely be discriminated by eye, if at all, but which possess chromosomal differences that make them reproductively incompatible. And most species that are each other's closest relatives fall, of course, somewhere between these two extremes.

This disconnection between speciation and morphological change has profound consequences for the evolutionary process. It means that two separate events, or at least mechanisms, are necessary for the "fixation" of evolutionary novelties. New morphological varieties first have to emerge, of course, and this is well documented to occur as a routine result of spontaneous genetic change—essentially, the introduction of "copying errors" as each individual's DNA is replicated for incorporation into his or her reproductive cells. It is these errors that produce the "mutations" whose histories have been exhaustively documented by geneticists, and most make little difference, though some have major consequences in individual development. In small populations, without other genes flowing in from outside, favorable mutations have a good chance of being incorporated as the local norm. But such new variants often have an ephemeral existence, since they are always at risk for reabsorption into the larger parental population should the barriers separating the parent and daughter populations break down. This is where speciation comes in. Should speciation intervene, and parent and daughter become incompatible, then the fate of whatever evolutionary novelties might be involved becomes inextricably tied up with the fate of the new species, no matter what external circumstances might change. And the new species itself assumes an independent historical existence, becoming available to be triaged by competition with other equally distinct kinds of organisms. Competition between species, indeed, makes as great a contribution to larger evolutionary patterns as does the competition among individuals within them. The question is simply one of the level—population, species—at which these actions take place.

Local Populations

Such larger factors add an additional layer of complexity to the evolutionary process, but they do not detract at all from the role of natural selection and adaptation. Wide-spread populations of any organism tend to diversify into local variants under these twin basic influences; and it is at this level of action that these "traditional values" come into their own. It is, indeed, small local populations, tied to particular local environments to which adaptation is possible, that constitute the "engines" of evolutionary innovation. Whole species can't usually do it, for they normally straddle several different environments, and adaptation is by its nature environment-specific. Some-times such innovation is small-scale; but occasionally it is major, which brings us back—briefly, I promise—to the question of "intermediates" in the fossil record. For it is a demonstrable fact that many very important innovations have appeared quite suddenly in the evolutionary record, unanticipated by earlier developments. This is the case, for instance, in the emergence of modern body form among humans. Primitive upright bipeds, with "archaic" body proportions (short legs and long arms, for instance), were around in Africa for millions of years before hominids of modern structure emerged. Yet throughout this long time the basic archaic body structure appears to have remained essentially unchanged, even as new species came and went; and when modern body form appeared, it did so suddenly, and pretty much out of the blue. The achievement of "modern" body architecture involved alterations throughout the skeleton, and a significant increase in body size as well; but all or most of this was achieved, it seems, in a single leap, rather than as a result of gradual fine-tuning over the millennia.

In his book *Sudden Origins,* my colleague Jeffrey Schwartz, rejecting like Eldredge and Gould the notion that in most cases we simply haven't yet found the intermediate stages, has recently elaborated an attractively simple mechanism by which many such sudden population changes may have come about. Schwartz points to a fairly recently recognized class of genes—the homeoboxes—that regulate major developmental patterns. Such genes are widely shared among animals, and small changes in the timing of their activities during the development of each new individual may result in differences as huge as some of those, say, between fruit flies and humans. Humans, for example, have eyes with single lenses that deform to produce focus, whereas fruit flies have multiple nondeformable lenses; yet even in these disparate animals the genes for eye development are basically the same. The difference occurs in the way in which the genes are switched on and off during development. Schwartz suggests that new forms of these regulatory genes originate by the same mechanisms as other genes, and thus that they are most likely to arise in the recessive state. Each new individual receives two copies (alleles) of each gene, one from each parent. These alleles may be dominant, or they may be recessive. A dominant allele will be expressed in the individual even if only one copy of it is present; recessives require two copies to be expressed. Thus newly emerged recessives will remain "silent" in the population until there are enough of them in the gene pool to make it likely that both parents will pass them along. At this point the anatomies they specify—potentially radically new in the case of homeoboxes—will appear abruptly in the population, with no prior warning. Here, then, is a mechanism whereby major anatomical novelties can suddenly arise within species. And once they are in place, of course, natural selection can take its course with them, whether positive or negative.

Adaptation

The matter of natural selection brings us back, inevitably, to that of adaptation, so closely are the two concepts intertwined. As I've already mentioned, the whole notion of adaptation sometimes comes close to being a tautology. It seems that almost any feature that has not been weeded out of a population—and many will not be, having decidedly neutral evolutionary effects—will be considered an adaptation by somebody. And whole schools of research have developed that involve the exegesis of one adaptation or another (the browridge, say, or the chin region). Wherein lies the problem. The longest-lasting legacy of the Synthesis turns out to have been the notion of transformation; and what's more, not so much the transformation of one species into another but the transformation over time of individual characters. Individuals are mind-bogglingly complex entities, and each contains a vast number of attributes that we might want to recognize as adaptations. But there has been a widespread tendency among evolutionary biologists to follow characters—adaptations—independently, as if they could properly be understood separately from the larger organisms of which they form part. And while it's fine to study, say, locomotion, it's clear that we will never fully understand how we got to be the upright creatures we are today without attention to the number and nature of those relatives who preceded us.

Yet often, such attention has not been forthcoming. For example, there is a whole coterie of paleoanthropologists who specialize in the evolution of the human brain, and indeed, at my home institution there is a very distinguished lecture series on this subject, in which I have myself participated. But when I was preparing my lecture a few years ago I decided to follow up on exactly what we knew about the evolutionary trajectory of the human brain. And I came up with a very dusty answer indeed. Yes, if we consider that over two million years ago, hominid brain sizes were very small (about great ape–sized), and that now they are three times larger relative to body size, we have to conclude that brain enlargement has been a major theme in human evolution. So far so good. But as soon as one tries to figure out the history of this enlargement, one runs into trouble. For a start, we don't know how many species we are sampling in the fossil record over the last two million years. We don't know what the range in brain size was in each one, and we don't know with any great accuracy what their time ranges were. Some well-known individual fossils are quite well dated—though others are not—but in no case do we have really reliable information as to the actual dates of origin and extinction of the species involved. This is not a great database for understanding what happened in the eventful period between *Australopithecus* and *Homo sapiens,* but we persist on proceeding as if it were. Certainly, in some way average hominid brain sizes have increased, quite dramatically in the aggregate, over the last two million years; but exactly how—which is surely what we really want to know—remains elusive.

In this case, at least, we are dealing with a phenomenon that has some minimal basis in empirical fact. Much of the discussion of "adaptations" doesn't even have this, as we will discover when we look at the arrogant pseudoscience of "evolutionary psychology" in a later essay. But for the moment it's enough to point out that each individual consists of a vast number of characteristics, some closely linked genetically among themselves and others not. And natural selection is a one-shot deal: Either individuals leave more than an average number of genetically related individuals behind them when they die, or they don't. What determines the reproductive fate of those offspring could be anything, but it will most likely be their overall viability. You can be as smart as you want, but it won't do you much good if you are too slow at escaping from predators. Your cryptic coloration may help you hide from your enemies, but it may not help in your breeding efforts. Your long tarsal bones may be advantageous in areas of small trees but hinder you where there is underbrush. In the end, natural selection has to vote up or down on the success—in living or in reproducing, or actually both—of the entire individual, not of its separate features. It cannot tease out subcomponents to favor or eliminate. There is no proportional representation here; the election is strictly first-past-the-post. Inevitably, then, we have to get away from the popular notion that evolutionary change consists essentially of a process of fine-tuning of individual characteristics. It simply cannot be. Individuals must succeed or fail—or reproduce successfully, or don't—warts and all. What's more, just as anatomical or cognitive features are embedded within individuals, individuals are embedded within the populations they belong to, and populations within the species of which they form part. The ultimate fate of evolutionary novelties depends on what happens at all of these levels. We will never be able to understand evolutionary histories without understanding the importance of this hierarchy, or without incorporating it into our historical accounts.

The bottom line, then, is that living creatures are not finely engineered devices, wherein one component after another may be removed for improvement, held up to the light, burnished, and replaced. They are complex organisms—genetically, anatomically, and ecologically—and they lead complex existences. What's more, they are just as subject to the vagaries of chance as they are to excellence of adaptation. After all, if your environment changes radically (as environments are wont to do), your vaunted adaptations may prove to be totally irrelevant. And this inevitably brings up the question of "exaptation."

No innovations ever arise *for* anything. They simply occur spontaneously, as the result of routine genetic processes. If they cause no problems, they can simply wait around for selection to do its thing, whether to their ultimate advantage or disadvantage—or neither. One aspect of all this is that very often novelties arise that either don't have any effect on the viability of the organism, or that serve usefully in a particular role that is different from the one in which they are ultimately co-opted. The classic example here is birds' feathers. These specialized features seem to have been around for a very long time as devices to help maintain body heat, long before being conscripted as adjuncts for flight. Equally, the peripheral speech mechanisms of human beings appear to have been in place for several hundred thousand years before they were conscripted for use in producing articulate sounds. So what might appear as an "adaptation" at one point in the career of a species might have initially shown up as an "adaptation" for something else, or for nothing much at all. Such "exaptation," it turns out, has been a major player in the evolution of many species. The role that chance plays in the evolutionary process should never be underestimated!

Diversity

My brief account of evolution here has not been intended as an exhaustive one, or anything approaching it. Evolution is a complex, multilayered process that functions on many levels, most of which we are still striving to understand fully. Still, I hope I have said enough to suggest why it is that the strongest signal we have yet been able to detect in the fossil record is that of diversity. If most people were challenged to give a thumbnail definition of evolution, of course, they would probably say something like "change in living beings over time." Well, yes, for the reasons we've already discussed, change is a powerful element in the overall evolutionary picture. Yet much more of the evolutionary history of individual species is characterized by stasis than by the incorporation of major innovations; and what is really important here is the *nature* of such change, and the context in which it occurs. Under the Synthesis change was viewed as slow, gradual, and virtually inevitable; but adding speciation and punctuated equilibria to the mix has changed our perceptions and expectations entirely. Fixation of evolutionary novelties relies on the production of new species, and appreciating this reality often depends on the abandonment of preconceived ideas.

This is often difficult, especially for evolutionary biologists, older generations of whom have had the Synthesis drilled into them from the very beginning of their careers. What's more, the notion of stately, gradual, linear change is itself actually quite a comforting one. It imparts, for instance, a sense almost of inevitability to the arrival of *Homo sapiens* on Earth; and in terms of figuring out where we fit into the overall picture of life, it certainly simplifies our task. Until, of course, we look at the details, which tell us that our species is merely one terminal twig—the surviving terminal twig—on a complex branching bush. This kind of scenario is a little bit more difficult to cope with than the traditional view. For example, I recently received an email from a colleague in China who inquired rather worriedly whether it was really true that I recognized seventeen species in the known human fossil record. And I had to do a quick count myself to confirm this number (since increased). Yet this multiplicity of hominid species over the past five million years simply brings our family into line with what we already know of the rest of the living world. From the beginning, the history of life has been one of continuous experimentation, of the production of new species and triage among them by competition with others, closely related or not. In this sense, extraordinary phenomenon though *Homo sapiens* is, we got here by perfectly ordinary means.

As time passes, we will come to have a better understanding of the complexities of both the evolutionary histories of individual organisms—including our own—and the evolutionary process itself. Yet those histories, and this process, have been so eventful, and have unfolded over such an extraordinary length of time, that it is vanishingly improbable that we will ever have a totally accurate picture of the biological background from which we emerged. No matter: Despite the talk of grand unifying theories it is not in the nature of science to explain absolutely everything. Moreover, scientific findings do not threaten anyone (except to the extent that *Homo sapiens* may prove incapable of controlling what science makes possible). But what it is critical to understand is that our species (or, for that matter, God) is not in the least diminished by the idea that we emerged thanks to the processes of evolution. The great deal that we have already learned serves only to emphasize Darwin's own eloquent conclusion, first articulated a century and a half ago in his *Origin of Species,* that there is "a grandeur in this view of life . . . having been originally breathed into a few forms, or into one; and that, whilst this planet has gone cycling on according to the fixed law of gravity, from so simple a beginning endless forms most beautiful and most wonderful have been, and are being, evolved."

Chapter Three

The Monkey in the Mirror

There are six billion human beings (for the moment); and no matter how bizarrely some of us behave or view the world, we usually have no great difficulty in interpreting each other's motives, or at least in explaining them away. But what about other species, even close relatives? Here we have to confess ourselves nonplussed, for it turns out that we are simply incapable of imagining states of consciousness other than our own. We know that we share the bulk of our evolutionary history with lemurs, monkeys, and apes. And we know equally that, as a result, we also share most or all of our brain structures with them. What's more, when we look into the eyes of an unhappy orangutan in a zoo, we share an instinctive feeling of kinship with it, and we *know* that it is feeling or thinking or experiencing something that we could relate to, if only we could figure out what it was. But precisely what is going on in the ape's head—or heart—will remain forever beyond us. It's not just a question of language, of the ape's verbal inability to explain. It's evident that the way in which even our closest relatives process information coming in from the outside world differs from our own. Just how large that difference is, though, and how it impacts the subjective consciousness of those relatives and others, remains one of biology's—and philosophy's—great conundrums.

None of this, of course, is to suggest that nonhuman animals lack "intelligence," or do not in some way possess "consciousness." Anyone who has lived with a dog, or even a parrot, or has spent any time observing virtually any kind of mammal in the wild, knows perfectly well that these animals are not automata. They are individuals, with their own perceptions of the world, and their own reactions to it. They are capable of anticipation, of pleasure, of fear, and of a host of other responses that are equally familiar to us human beings. The fact that many other organisms clearly have so much in common with humans has, in fact, led some to conclude that what's most important to understand is that which unites rather than divides us. On the other hand, others feel strongly that it's human uniqueness which is necessary to define. In reality, though, the two interests are effectively identical, since we can't understand exactly how unique we are without knowing the full extent of what we share with our closest relatives. As I've already implied, this is a tough job indeed; but at least it's possible to look at some of the aspects of the behavior of monkeys, apes, and other mammals, and to see how closely they resemble our own. What this means in terms of these relatives' subjective consciousness and experience of the world is, of course, a different matter.

Intention and Deception

Dissecting out the aspects of consciousness is a tricky task, since awareness is multidimensional—and who knows, in this rich stew, what is cause and what is effect? Since we have to start somewhere, though, let's begin with the matter of intention. Everybody can agree that a major aspect of consciousness is the ability to form intentions; and nobody will dispute that human beings spend much of their lives in this activity, however hollow those intentions may eventually turn out to be. So much for humans; but what of other mammals? Could anyone, for example, reject the notion that when your dog runs up to you with his ball, drops it at your feet, and looks expectant, he does so with the intention or at least the hope that you will throw it for him? Well, yes, some hard-core skeptics could, claiming with some justice that we should seek explanations for such apparently complex behaviors at the simplest possible level, and mumbling words such as "reinforcement." But why should we not conclude that in this case, at least, humans and nonhumans might have similar motivations? After all, they share the same basic brain structure; and whatever it is that nonhuman mammals experience, it has to be through essentially the same electrochemical phenomena that drive our own mental functioning.

Actually, some philosophers would dispute even this, arguing that mental functions and brain events are not strictly comparable because the former are entirely subjective while the latter are objective. But this is a philosophical argument, not a biological one; and as means improve of imaging the human brain as it carries out well-defined tasks, the link between the two categories becomes steadily clearer—as also, it must be admitted, does its complexity. Not that the subjective nature of consciousness and its components fails to pose problems for scientists as well as for philosophers. Scientists, after all, are a lot happier studying phenomena that are more easily isolated, described, and measured than are the elusive notions of human and nonhuman awareness. And of course, if consciousness were something more susceptible to scientific analysis than it is, we would certainly know a lot more about it by now than we do—which is very little indeed. Nonetheless, to return to the original proposition, it seems like hair-splitting to deny nonhuman mammals some capacity for forming intentions—and even for insisting on them, as your dog undoubtedly will, dropping his ball in front of you time and again until he gets your attention.

A more complex level of cognitive function is represented by deception, for in its strictest sense this activity involves not only intention (to deceive), but also an understanding of what is going on in the mind of another as well as in one's own. To deceive successfully in the human fashion you must possess an appreciation of your victim's mental processes and adjust your own behavior accordingly. Primatologists have looked quite extensively at deceptive behaviors in both wild and captive nonhuman primates, and have had little trouble in finding examples of this rather unattractive propensity. Of course, it can always be objected that some forms of nonhuman deception simply involve whatever works in a given set of circumstances, rather than a fully formed intention to deceive. One such example might be that of an infant monkey who finds himself next to another group member whose food he covets. By screaming as if attacked, he attracts his mother to drive his neighbor away, and thus gains access to the food. The parent is certainly deceived; but is this ploy actually no more than something that the infant has simply found to be effective when he is hungry, in which case there would have been no intent to deceive as such? Or is there a preconceived strategy here? You could argue this particular instance either way, but there are other examples among nonhuman primates in which the matter of intention is less ambiguous.

The best instance of the latter, which involves both bluff and double-bluff, comes from a study of captive chimpanzees who were held in a large enclosure. One member of the group (Belle) was regularly shown where food had been hidden, and at the beginning of the experiment she duly passed the information along. Soon, however, a young male (Rock) began to monopolize the food when its whereabouts had been revealed, prompting Belle not to show the others where it was when Rock was around. Instead, she sat on the cache. Rock, the stronger of the two, responded to this by pushing Belle aside to get access to the food, whereupon she took to sitting somewhere else, and not visiting the food until he was out of sight. Rock then upped the ante by looking or moving away until Belle made a move for the food, which he then grabbed. Belle's response was to lead the whole group in the wrong direction, and then to rush back to the cache while Rock was uselessly foraging. But it didn't take Rock long to catch on to this ruse too. Sometimes a single piece of food was hidden apart from the main pile, and Belle would then lead the group to it, sneaking off to the larger food source while Rock was busy eating the single piece. Before long, though, Rock realized what was going on, and started to ignore single food items. By this stage, Belle was out of options, and was reduced to staging temper tantrums.

Many of the behaviors in the complex saga of Belle and Rock might be explained by conditioned learning; but by the time we find Rock feigning a lack of interest in order to sucker Belle we are surely correct to conclude that he had understood the logic of her actions, and was anticipating her next move. He had, to put it another way, begun to form an image in his own mind of what was going on in Belle's. Various other reports have confirmed that this is a fairly routine ability among great apes. But, significantly, there are as yet no indications that monkeys can do the same thing.

The Monkey in the Mirror

Somewhat related to the issue of whether individuals can read the minds of others, but distinct and even more difficult to approach experimentally, is the question of self. Human beings have no difficulty with the concept of self; and it can, indeed, be argued that our only true knowledge is of ourselves, since our perceptions of the outside world are filtered through an elaborate and highly subjective cognitive apparatus. I have already said that nonhuman mammals are far from being automatons, and this is clearly true; but does it necessarily follow that they have a concept of self that would be broadly familiar to us? The answer to this is almost certainly no; but it has to be admitted that the degree to which nonhuman primates may or may not have an internal image of self is a devilishly hard question to approach. Essentially, we have to proceed by approximation. While self-knowledge is surely something distinct from self-recognition, so far the best that cognitive science has been able to do in this realm is to use the rather crude tactic of employing mirrors to see whether individuals can recognize their own reflections.

It should be noted that even human beings who are not conversant with the qualities of mirrors experience a period of confusion when presented with them for the first time. It's thus hardly surprising that great apes take a while to catch on. However, once habituated, most display a ready ability to recognize themselves. The most dramatic demonstration of this propensity involved anesthetizing chimpanzees and orangutans, and painting spots on their faces while they were unconscious. When they recovered, most of the subjects wandered around nonchalantly until they caught sight of themselves in a mirror. Their immediate reaction was to use the mirror as an aid in picking the paint off their faces. Clearly they had recognized themselves, and they were soon pulling faces and exploring their persons using this unfamiliar opportunity. Interestingly, several gorillas tested did not seem to recognize themselves, although one, the famous Koko, a sign language star, definitely does recognize her own reflection.

So far so good, perhaps; but does the ability to recognize oneself in a mirror convincingly demonstrate that one has a *concept* of self? This is a tough issue, but most cognitive scientists would, I think, argue that without such a concept individuals would lack any means for interpreting the reflected image, and would thus be unable to recognize themselves. Nonetheless, even if we accept this, where does it leave us? It seems equally likely that recognizing one's reflection is only a part—maybe, even, just one small consequence—of what we human beings are familiar with as the concept of self. Which inevitably brings us back to the essential problem of defining human consciousness, and of our inability to imagine the subjective experience of forms of awareness that are not fully like our own.

Still, the fact that most apes recognize their own reflections in mirrors surely is significant at some level, especially when we realize that monkeys do not. Monkeys are clever creatures, and they are adept at exploiting the qualities of mirrors for useful ends. There is, indeed, plenty of evidence that Old World monkeys have a keen sense of visual geometry (in fact, some of the most striking documented cases of visual deception are based on monkeys' appreciation that different things—or individuals—are visible from different angles). Given this ability, it is hardly remarkable that captive monkeys will spontaneously use mirrors provided by researchers to see around corners and to identify other individuals lurking there. But comfortable as monkeys may become with mirrors and their properties, it has also been shown that they cannot identify their own reflection in a mirror. Instead, they react to these images as though they were seeing strangers. A monkey that has had a mark painted on a visible part of its body under anesthesia will try to remove it; but if the mark is painted on its face, it will make no such attempt even when provided with a large mirror.

What do we make of all this? First, it is evident that there is a qualitative difference among the perceptions of self exhibited by monkeys, apes, and human beings. Whether we can arrange these perceptions into any kind of a historical sequence is, of course, dubious. Each of these primate types represents an end point of a long, independent evolutionary history; and to regard monkeys as junior-league apes, and apes as inferior humans, is deeply misleading. The most that we can say, if we are brave enough to say anything at all, is that we have here a functional spectrum: Perception of self is more finely honed among human beings than it is among apes, and both outshine monkeys in this regard. The same is also presumably true of intentionality; it is certainly true in the case of deception; and it is very emphatically true in terms of the general complexity of social life and diverse other behaviors. Now, perhaps all that I am saying here is that humans are smarter than apes, which in turn are smarter than monkeys. This statement has the advantage that nobody would wish to quarrel with it (and how many statements could you say that about?). But it really doesn't get us very far. Intelligence is a complex and multidimensional attribute. To say that a person (or an ape, for that matter) is smart doesn't tell us anything very specific about him; it certainly conveys much less than saying that someone is good at programming computers, or at fixing TVs, or has a good memory. So what do we mean by intelligence? It's worth taking a quick closer look.

Brains, Intelligence, and Consciousness

Two million years ago our precursors had small brains, both in absolute size and in relation to the size of their bodies. Today our brains are three times larger. What could be more natural than to conclude that in the intervening period hominids have become more intelligent? Well, yes, in the most general of senses this must be true.

There is no doubt that our intelligence comes ultimately from our brains; and it's equally well established that the brain is a very expensive organ, accounting for a disproportionate amount of the body's energy consumption. You don't develop a big brain for the fun of it, for it simply costs too much; and the payoff has somehow to lie in advantages that we suppose must be related to behavioral improvements. But we know remarkably little about the actual sequence of events in human brain enlargement over time. Even less do we understand the effects of these events. We tend, of course, to think in a very linear fashion about brain expansion. Logically enough, we conclude that every little bit of extra brain equals a little bit more intelligence. And if a certain quantity of intelligence is good, wouldn't a little more be better? Well, yes and no. Other things being equal, it is presumably better to be smarter than dumber. But brains are amazingly complex structures, and adding or reconnecting a bit more neural matter here or there could have a whole range of consequences. Intuitively, from a human vantage point, it's hard to avoid the conclusion that, somehow, brain expansion is intrinsically a good thing—though perhaps the contemplation of the extreme rarity of this phenomenon in nature should make us think again. What's more, such notions tend to skate over the complex realities of the evolutionary process;

and, as it turns out, the concept of a gradual increase in brain size over the eons is actually rather problematic.

For a start, this idea strongly implies that every ounce of extra brain matter is equivalent in intelligence production to every other ounce—which is clearly not the case. Although brain size is clearly important at some level, what almost everyone acknowledges is that the critical factor is brain *organization.* The big problem here, though, is that we don't have any fossil brains. We have only fossil brain *casts*—impressions of the outside form of the brain. And paleoneurologists, the folks who study such things, tend not to agree on what external brain form is actually telling us about the functional capabilities of these organs. Which leaves us with not much more than raw brain size to consider. And here we are hardly helped by the fact that today the human brain is probably more variable in relative size than is any other major soft tissue structure, ranging in volume from around one to two liters. What one is supposed to make of this is unclear, since nobody has yet contrived to show that among living people there is any correlation at all between brain size and levels of general achievement.

In the fossil record we have additional problems. For instance, we don't have many good fossil brain casts, and often we have yet to solve the problem of which species those we do have actually belong to. But even when we think we know those species, we often don't know how old the specimens that represent them are. And even if we have a reasonably good handle on the age of the specimens, we usually don't have any idea at all what the longevity was of the species they belonged to. Did this specimen or that come from the beginning of a species' tenure on Earth, or the middle, or the end? And, speaking of ranges, what was the variation in brain size within the various species? Does a given brain cast represent the top end of the range, or the bottom, or something in between? Given all these uncertainties, you can see why I'm reluctant to say anything at all about the pattern of hominid brain expansion over the past couple million years, except to agree that at the beginning of this period the average hominid brain was small, and now it is large. Of course this *must* mean something significant, but exactly what that something is, is another question.

Well, if fossil brains aren't yet the key to understanding the emergence of our astonishing human cognitive powers, is there an alternative approach? Lots of people have tried other tacks, and despite its elusive characteristics, human intelligence has elicited a huge amount of discussion. Intuitively, of course, we all know what intelligence is, although definitionally it's often hard to get far beyond such tautologies as "intelligence is whatever it is that allows one to behave intelligently."

Clearly, high intelligence involves some quality of flexibility in response to problems coming in from the environment. It's not just an ability to respond in some way to stimuli: All organisms have that. And, equally clearly, some degree of behavioral flexibility is not a uniquely human ability. But when we're studying living animals it's sometimes tough to separate out intelligent behaviors from those that just *look* intelligent. You can, after all, get the right answer for the wrong reason. Cognitive scientists tie themselves into knots trying to determine whether behaviors they observe in chimpanzees and other nonhuman primates result from ratiocinative processes, or simply reflect conditioning, or imitation, or some other factor.

What's more, as computers become more sophisticated, we face this problem even in regard to inanimate objects. A half-century ago, the great computer theorist Alan Turing famously proposed a theoretical test for determining whether a machine could ever think. You're probably familiar with it. Turing's idea involved an exchange of messages and questions between a machine and a human operator in different rooms. If, at the end of a fixed period of time, the operator was unable to tell whether he or she had been interrogating another human being or a machine, the conclusion would have to be that the machine was capable of thought. Turing believed that such a machine was not far beyond the horizon; but in fact, despite all the effort that has lately been poured into artificial intelligence studies, this test has not yet been passed. Even powerful computers that can beat the best human chess players don't do the trick. Why? Well, computers by their nature are algorithmic, rapidly applying a fixed set of rules to the solution of well-defined questions. This is, of course, an ideal approach to the kinds of problems that arise in a chess game. But one thing nobody would wish to dispute is that what goes on inside IBM's Deep Blue and in the mind of chess player Garry Kasparov are quite different things.

As computer science progresses, and computers become more adept at "learning" from the results of their own labors, a greater convergence may be achieved between what it is possible for a computer and a human being to do (using computers!). But it is vanishingly improbable that any machine will ever achieve an internal state that we would recognize as consciousness. Partly this is because human consciousness is very clearly nonalgorithmic. Even if we have to conclude, for want of a better possibility, that subjective awareness in ourselves and our relatives is due ultimately to a mass of electrochemical events in the brain (just as calculation is due to analogous events in a computer), it is equally clearly not the product of a mechanism that dutifully clicks through a listing of tasks and "if-then" choices. The human brain is the outcome of a hugely long evolutionary history, being descended from simple structures that first emerged something like 400 million years ago. Since the appearance of that proto-brain in the earliest vertebrates, this organ has grown by accretion, from time to time adding a structure here, expanding one there, and acquiring a new connection somewhere else. On the other hand, as far as is known, not much if anything has been "lost" in the course of human brain evolution. Our skulls still house the descendants of structures that eons ago governed the behavior of ancient fish, of primitive mammals, and of early primates. In particular, the pathways via which messages are transferred from one newer brain region to another travel via some very ancient structures indeed (structures that principally mediate emotional response). And such messages may well be affected by their journey. The accretionary history of our brains has assured that they are not the kinds of machine that any rational engineer would design from the ground up; and it is, indeed, very likely to this untidy and adventitious history that our vaunted cognitive capacities are due.

The neurobiologist Antonio Damasio has recently devoted most of a book to the proposition that the remarkable human cognitive capacity exploits this long history. His basic functional point is that consciousness is representational, depending on the making of associations between objects and the self (presumably, as in the case of our monkey, whether fully realized as such or not). More importantly, though, Damasio argues that such associations are mediated not by thought but by feeling. He further believes that consciousness comes in two forms. First is "core consciousness," which is limited to the here and now, and is what we share with other higher primates. The ingredient we have that makes us unique, he calls "extended consciousness." This type of consciousness adds awareness of past and future to the mix. Well, as with everything else in this intractable subject, it is arguable whether this is a totally accurate (or at least complete) categorization of what we subjectively experience as consciousness (think of your dog running to the front door when you take out your car keys). But Damasio is surely right when he points to the intimate link between higher cognitive functions and emotional processes in the brain. Little as we understand the highly complex workings of our brains in producing consciousness, it is clear that there is a "whole brain" effect in the production of our prized awareness. Yes, it is now possible to identify particular brain areas as principally involved in the performance of specific tasks; but when it comes to our generalized conscious awareness, in particular, we almost certainly have to look at how the brain functions as a whole—new brain, old brain, the entire thing. There are,

I'm pretty sure, two components to reasoning: symbolic and intuitive. And the intuitive processes, in particular, depend on some very ancient brain structures indeed. Almost certainly, it is the combination of symbolic reasoning with intuition that gives human beings their remarkable creativity; and there seems to be little point in looking for some identifiable nucleus of brain cells whose presence or absence governs whether or not its possessor will possess consciousness or great intelligence. We should, in other words, leave Einstein's brain in peace. Of course, as techniques for imaging the brain activity of living subjects improve, we are daily learning more and more about specific brain functions; but the enveloping questions of consciousness and of our ability to exhibit "intelligent" behaviors remain elusive. They will almost certainly be answered (if they can be: Are we intelligent enough to understand the roots of our own intelligence?) on a higher level of brain integration.

Types of Intelligence

One of the most important things to appreciate about intelligence is that it is not a unitary quality. Indeed, intelligence is largely situational. It's not entirely unfair to claim that those people with high IQs are simply those who are best at doing IQ tests, though it may also be true to say that people who score high in one category are more likely to do the same in others. What is uncontestable, however, is that some people are better at doing specified mental tasks than others are, and that some do more such tasks well than others do. Some people are hopeless at certain operations, such as programming the VCR; but the same person might be a brilliant cook, or artist, or rat-catcher. Then, of course, there is the idiot savant, who can perform astonishing mental calculations while being totally out of his depth in the social milieu. Nobody is outstanding at everything, and few people without major cognitive deficits fail to show what we would recognize as intelligence in some form—and all are sentient beings. This is important, for the fact that intelligence is so clearly not a generalized ability but is rather something that expresses itself in a variety of apparently unrelated ways, has led some scholars to a remarkable conclusion. This is that, in some fashion, intelligences, and by extension the brain that lies behind them, are "modular." These folk claim, in other words, that there are qualitatively different types of intelligence; and if this is true, it leads one to ask whether those intelligences were acquired as a unit, or at different times. Is the human brain like a general-purpose computer that has become more powerful over the millennia by adding more memory and connections? Or is it, instead, an agglomeration of distinct mechanisms that have been added one by one, just as the basic Swiss army knife inexorably acquires more and more blades? Or, alternatively, is it neither?

Some scholars, notably those of sociobiological inclination, have identified a huge number of individual mental modules. There are modules for recognizing faces, for friendships, for social exchange, for kin recognition, for fear, for child care, for a theory of self, and on and on—you get the idea. Yet what this really adds up to is little more than a laundry list of activities in which human beings (and, for the most part, other species too) indulge. This listing may be of interest to those bent on sociobiological explanation, but it is otherwise not very useful at all in understanding how the brain puts its perceptions together to produce thought and intelligent response. A more interesting, if related, tack has been taken by the archaeologist Steven Mithen, who has analogized the evolution of the mind to the building of a cathedral. The "nave" is composed of a module of generalized intelligence, with doors opening to modules concerned with perception of stimuli coming in from the outside world. Next are added multiple "chapels" devoted to individual cognitive domains, which he identifies as "technical," "social," "natural history," and possibly "linguistic" intelligences. These modules of intelligence connect only to the nave, and cannot communicate with each other. Finally, though, doors are opened between the chapels, permitting a "cognitive fluidity" to ooze forth. All this is very arguable, but Mithen analogizes the first stage of cathedral construction to the capabilities of chimpanzees, with the elaboration of general and social intelligences. Stage 2 sees the origin of toolmaking and meat-eating (i.e., technical and natural history intelligences, and maybe incipient "linguistic" intelligence). The final act of construction throws open the portals between all of these modules, allowing symbolic associations to be made.

It is not Mithen's fault that at the present state of our knowledge virtually any presumptive account of the evolution of human cognition almost inevitably has to be presented in the form of a "just-so" story of this kind. But his scenario presents all the problems of the "fine-tuning" notion of evolution that we discussed in the last essay. Not only does it leave out the emergent qualities of human cognition, a subject to which we'll return, but it also ignores the fact that not all of the components of each of the putative intelligence modules are very closely correlated, at least to judge from people today. And it does nothing very much to help explain why most apes can recognize themselves in mirrors, while monkeys can't. Unless, of course, we invoke yet another module, for self-recognition; but this would ultimately do nothing more than simply lengthen our laundry list. What's more, while it is all very well to dissect out intelligences of various kinds to exegecize, as far as brains (and, come to that, living behaviors) are concerned, we only have our own and those of some pretty remote relatives to study (humans and great apes last shared a common ancestor at best five to six million years ago). Some behaviors are fractionally less problematic, since the archaeological record does produce a shadowy representation of what our own precursors actually did, if not thought, at least in some realms of their living existences. Meanwhile, there is one uncontrovertible thing that we can say about the uniqueness of our own consciousness. This is that, while every other organism we know about lives in the world as presented to them by Nature, human beings live in a world that they consciously symbolize and re-create in their own minds. Which is what makes us such fascinating—and dangerous—creatures.

Chapter Four

Human Evolution and the Art of Climbing Trees

Some 1.6 million years (1.6 myr) ago, an adolescent boy, perhaps nine years old, stumbled into a swamp on the west side of what is now northern Kenya's Lake Turkana. Whether he was sick, or lost, or abandoned, can be debated; but whatever the case, sadly this child died, presumably alone, in this uncomfortable and insect-ridden place. Just another small tragedy, you might think, in the long history of such tragedies that has given us the human fossil record. For paleoanthropologists nearly two million years later, however, this sorry event gave rise to a true bonanza. The boy's body was not immediately found and dismembered by scavengers, as is the fate of almost every dead creature on the African plains. Instead, it floated face down for a while until it was covered by swamp muck and its bones thus preserved for posterity to discover. For science, what an event! The human fossil record largely consists of fragments of skull and body bones, from which it is often very difficult to reconstruct the lives of their original possessors. A nearly entire skeleton (missing essentially just the hands and feet) is an almost unheard-of paleontological luxury. Indeed, prior to the remarkable discovery of the Turkana Boy, in the mid-1980s, only one truly ancient and tolerably complete human skeleton was known. This was the famous 3.2-myr-old Lucy, discovered in Ethiopia ten years earlier, and far less whole than the Boy. Lucy caused quite a stir, though, for in her day this tiny and very primitive creature was the only respectably preserved hominid skeleton from the entire period before the Neanderthals began to bury their dead around 50,000–100,000 years (50–100 kyr) ago. So the Turkana Boy filled a huge void in our knowledge of the human evolutionary story, and paleoanthropologists were justifiably entranced by this extraordinary find.

But perhaps we should not get too far ahead of ourselves in the story of how our precursors became human. Lucy and the Turkana Boy are certainly the stars of this drama, but the cast of characters is a large one, and the action began at least a million years or two before before Lucy's time. So let's start at the beginning, both in terms of the human fossil record, and in terms of the history of the ideas that record has elicited. Five million years ago our precursors on this Earth were a very different lot from us. Like Lucy, they were small-bodied and small-brained, and they retained a lot of characteristics that were useful for moving around in trees. Now we are tall, big-brained, and committed to life on the ground. Surely, then, what happened in between must have been a matter of becoming bigger, enlarging our brains, and abandoning our ancestral arboreal adaptations. Well, yes, in a very limited sense I suppose that this does sum up the whole thing. But what does it actually explain? Not much, and the more you think about it, the more evident it becomes that the really interesting question is *how* all this came about. The hard part here, unfortunately, is that beginning to sort out what actually happened in human evolution necessitates abandoning a lot of comforting assumptions about how the evolutionary process itself works. In making the ultimate emergence of *Homo sapiens* seem almost inevitable, received evolutionary wisdom leaves us with relatively little to be explained.

Thanks to the overwhelming triumph of the Evolutionary Synthesis in the years following World War II, human evolution, like that of other organisms, came to be seen as a gradual, linear process that, come hell or high water, continued doggedly along a path of inexorable betterment. Ever vigilant, natural selection would assure it. And since the admittedly impressive (if also appalling) hominid of today appears to itself to be so manifestly an improvement over its predecessors, it was virtually a foregone conclusion that, given enough time, it was going to come out on top. As the fossil record enlarged, though, it became evident that the history of our species has actually been a good deal more complicated than a single drawn-out event of transformation. It has taken a while for this reality to sink in, and there are still paleoanthropologists who resist the notion that the story of hominid evolution is dotted with failures and blind alleys. But ultimately the conclusion is inevitable: We did not get here as the result of a process of gradual refinement. The reality is quite a bit more eventful than that, and it's certainly more interesting than a simple morality tale about evolutionary virtue rewarded.

In this essay I'm going to look at the pattern of appearance of what all would agree are the three most momentous (in a sense, nontrivial) innovations in the hominid record prior to the emergence of *Homo sapiens*. The first of these is the appearance of upright posture, underpinned by several very specific anatomical structures, as best seen in Lucy. The second is the introduction of stone tool technology and the cognitive innovations that must have accompanied it. And the third such innovation is the acquisition of modern body form—dramatically exemplified by the Turkana Boy—which also constituted a spectacular break with the past. Nothing we know of in the hominid record prior to our own appearance has had such profound consequences as bodily modernity; and even though the record may be a lot more opaque than we would like, it is clear that we are not looking here at a linear transformation.

The twentieth century witnessed endless debate about the nature of the original hominid adaptation: the innovation that placed us apart from the forest-living ancestors of our closest relatives, the apes, and set us on course for becoming human. At one stage, big brains were given prominence in this process; and other possibilities championed over the years included clever hands and a variety of other advances that it was believed had ultimately led to toolmaking. Today's leading candidate, however, is the adoption—in some form—of upright bipedalism: which is to say, locomotion, or at least movement, on two legs. It's important to bear this last distinction in mind; for while paleoanthropologists nowadays are uncharacteristically unanimous in believing that standing up in some form lay behind the origin of our family Hominidae some 6 to 7 myr ago, exactly what went on in this process is a matter of considerable argument.

What kind of stimulus made it happen is also a matter of controversy, though evidence is accumulating that for millions of years prior to the emergence of hominids, the world's climate had been slowly cooling and drying. This culminated, between about 6 and 5 myr ago, in the Terminal Miocene Event, which saw a diminution in the area of Africa that was under forest cover. As the forests shrank, of course, forest-edge environments (probably the preferred habitats of the earliest hominids) expanded, and the African fauna took on a new aspect as open-country antelopes, for example, began to proliferate at the expense of forest browsers. At about 2.5 myr ago, there followed another episode of climatic deterioration, which saw yet more forest fragmentation and faunal replacement. Finally, another cooling episode took place at about 1.8 myr ago (ushering in the Ice Ages glacial cycle in which we still find ourselves today, with the Earth's climate oscillating from cold to warm about every 100,000 years). However, since the African fauna had by then been "winterized," the effects of the 1.8-myr event on the animal populations with which early hominids coexisted were less dramatic than earlier ones. Interestingly, though, each of these climatic episodes broadly coincided with one of the major events in human evolutionary history that we will look at below.

The First Bipeds

The earliest pretenders to hominid status are both over 4 myr old, and both come from sites in eastern Africa.

The earlier of these, rejoicing in the name of *Ardipithecus ramidus,* is from the 4.4-myr-old site of Aramis, in Ethiopia, and consists of some very fragmentary fossils described in 1994, plus some more complete remains found since then that are proving nightmarishly difficult to separate from the matrix that encloses them. On the basis of some very indirect evidence, it was initially suggested that *A. ramidus* had been bipedal; a more recent suggestion by its discoverer is that it "had a type of locomotion unlike anything living today. If you want to find something that walked like it did, you might try the bar in *Star Wars.*" Indeed, it's not yet entirely clear that *A. ramidus* was actually a hominid; but if it was, at present its major importance is in demonstrating that, from the very beginning, the hominid picture was one of diversity. *A. ramidus* is very different from the next-in-line hominid species, which is very little younger; and there is no question at all of an ancestor-descendant relationship between the two. Evidently, right from the start, our precursors have ramified to explore the manifold different ways that there are to be hominid.

Somewhat better documented than *A. ramidus* is the species *Australopithecus anamensis,* known from sites in northern Kenya that date from about 4.2 to 4.1 myr ago. While a bit different in the construction of its jaws and teeth from the best-documented early hominid, *Australopithecus afarensis,* the new form from Kenya seemed to fit fairly comfortably into the same general group (the australopiths)—though the assortment of fossils attributed to it is motley enough to suggest that it would benefit from another look. In any event, what really caused a stir when *A. anamensis* was announced was that among the elements attributed to it was a shin bone that bore unequivocal evidence of bipedality, especially in the way in which it contributed to the knee joint. Since this discovery, in 1995, nobody has disputed that some form of uprightness had been incorporated into the hominid behavioral repertoire by about 4.2 myr ago. But exactly what form?

To answer this question, it is useful to touch on the rather contorted history of interpretation of *A. afarensis.* This species, first described in the 1970s from sites in Ethiopia and Tanzania, and now believed to span the period between about 3.8 and 3.0 myr ago, is by a country mile the best-known of the early hominids. This is due partly to the miraculous finding of our friend Lucy, the 40-percent-complete skeleton of a tiny individual dating from about 3.2 myr ago; partly to the astonishing "First Family," a slightly earlier assemblage of broken bones that may represent the remains of an entire group of australopiths that died in a single cataclysmic event, perhaps a flash flood; and also to a raft of other fragmentary fossils patiently sought after and found over the decades, including a recently recovered skull. These fossils are also complemented by the miraculously preserved footprint trails from Laetoli, some 3.5 myr old.

Finds like Lucy and the Laetoli prints are extraordinarily rare in the human fossil record, and are thus a precious resource indeed; but of course it would have been too much to hope that, because of the abundance of evidence they provide, unanimity would have been rapidly achieved in their interpretation. There is, indeed, still some question as to whether only one species is represented in the large collection of *A. afarensis* fossils known, though most current controversy is in other areas.

The greater part of this controversy concerns the degree of bipedal adaptation of these early hominids. First reports of Lucy and related specimens pointed almost exclusively to humanlike features, particularly in the pelvis and knee joint. Here, it was concluded, was evidence of the establishment of an essentially modern form of bipedalism very early in human history. Similarly, early studies of the Laetoli footprint trails stressed their basically human character. But it wasn't long before various scientists began to point out anatomical characteristics in *A. afarensis* that were not concordant with a fully humanlike pattern of locomotion. Various aspects of the knee and upper limb were first called into question; but soon there was a small industry devoted to showing that, for example, the long, curved hand and feet bones, the conical trunk, and the short hind limbs of *afarensis* would not only have been a handicap when moving on the ground, but would also have been an asset when feeding and moving in the trees. Indeed, it was pointed out that trees might well have provided the only relatively safe sleeping places for hominids, whose distributions were probably limited by the availability of such sites. At the same time, new studies of the Laetoli tracks began to emphasize their apelike characteristics, and some studies of living chimpanzees pointed to significant amounts of bipedal behavior while the apes were moving in the trees.

The argument continues; but perhaps the outlines of a scenario of early human locomotion, at least as exemplified by *afarensis,* are beginning to emerge. The structure of the pelvis, just by itself, indicates that a major reorganization of the skeleton had occurred. The pelvic structure of *afarensis* is that of an essentially upright creature that needed to support the abdominal organs that now pressed down on the pelvis. So far, so good. But does this mean that *afarensis* walked upright in the way that we do? Here the answer is certainly negative. Of course, any functional inferences from bone form are just that—inferences. However, among all the noisy competing claims, one thing is pretty sure: Lucy and her kind did not locomote in anything like the modern human fashion.

They may have moved bipedally when on the ground, at least some of the time—the Laetoli prints alone are good enough evidence that they did—but it's not necessary to conclude that they used only upright postures when they were on the ground, or that the bipedal features of their body skeletons were necessarily "adaptations" to terrestrial life. Although the structures of the hands and feet and numerous other features of these hominids were clearly inherited from an arboreal ancestor (hardly a surprise), this doesn't mean that they were simply "evolutionary baggage" that had to be lugged around inconveniently. If australopiths possessed these features, then whatever abilities came along with them were undoubtedly part of their behavioral repertoire, a repertoire that allowed them to live comfortably in the forests and along the forest edges, as well as to venture at least occasionally (and probably nervously) into open environments, such as the one represented at Laetoli.

What's more, australopith locomotion as we see it reflected in Lucy was not simply a transitional stage between apelike and humanlike modes of getting around. If we didn't know this from anatomical details, we would still know it from the long-term apparent stability of this particular body form. Lucy does not represent an ephemeral waypoint along the journey from ape to human; rather, her structure reflects what was apparently a highly successful and stable adaptation in its own right. From the very first hints of bipedalism in *anamensis,* right through to the last appearance of australopiths more than two million years later, basic body build stayed, as far as we know (and we wish we knew more), pretty much the same. Of course, the australopiths were a diverse group. Not only did they branch into the two major and well-documented lineages that are familiar as the big-toothed "robusts" and the lighter-skulled "graciles," but they diversified throughout the African continent into a substantial variety of species (at least nine are currently recognized even by conservative paleoanthropologists, with more to come). As far as can be told, once the initial breakthrough in body form had been made, differences among species remained relatively minor. This is so even though the basic division into gracile and robust lineages, for instance, is thought to have involved a significant divergence in dietary patterns. The bottom line is that, while there were variations in abundance, the basic theme persisted. And this in itself reflects a larger pattern that consistently crops up throughout the history of human evolution: Despite the strong signal of variety at the species level, right up to the appearance of *Homo sapiens,* major innovation in human evolution has been sporadic indeed. What's more, the expected motif of refinement, of fine-tuning, of an inexorable progression toward ourselves, is conspicuous by its absence. Species come and species go; and, once in a long while, something unexpected emerges. Finally, tempting though it is to "explain" behavioral novelties by the arrival of new species, it is actually very difficult to pin down innovation in the technical realm to the appearance of new kinds of hominid.

The Earliest Toolmakers

Stone tools constitute the earliest direct evidence we have of early hominid behavior. In the case of all other extinct organisms, what we know of how they behaved has to be inferred from bodily structure or from other forms of indirect evidence such as isotope ratios believed to reflect diet. In both instances, these are pretty rough indices of actual comportment. Once stone tools began to be made and used, however, our knowledge of hominids acquired a unique historical dimension: the archaeological record. There is an unprecedented profundity to this kind of archive, because not only do the durable stone tools themselves survive the ravages of time, but so also does evidence of how they were used. Not, of course, that this greater depth of knowledge necessarily renders interpretation any easier; archaeologists have at least as much difficulty in achieving unanimity as do the biologists.

In the case of the origin of the archaeological record itself, the first problem occurs in determining where it actually starts. Crude stone tools were first made by bashing lumps of stone with other lumps of stone, a process that doesn't take place only in the hands of hominids but happens all the time on the bed of every river. Long before true early tools were discovered in Africa, so-called *eoliths* ("dawn stones") had puzzled antiquarians in Europe. These were small lumps of rock showing breakage due to impact, but they eventually turned out to be products of nature. The difficulty that still faces archaeologists, though, is that the first stone tools probably didn't look very different from eoliths: small cobbles with a flake or two knocked off them by a few blows. And by the time more elaborate and thus more easily recognizable tools were made, an indeterminate amount of time had elapsed.

The first generally recognized stone tools come from sites in eastern Africa that date to about 2.5 myr ago, or perhaps a little more. They are generally similar to those first identified at Tanzania's Olduvai Gorge, and are thus known as "Oldowan." They consist of small cobbles, only a few inches in length, with one or more sharp flakes removed from them by percussion. Although the first such tools to be identified consisted of the "cores" from which the flakes came, archaeologists believe it was actually the flakes that were used as cutting tools, while cores were often used for pounding. And whereas a small sliver of rock an inch or two long doesn't sound like much, the flakes turn out to have been surprisingly efficient utensils, which have been used by experimental archaeologists to butcher entire elephants!

One of the uses—perhaps the principal use—to which these simple tools were put was to butcher the carcasses of dead animals. Archaeologists have argued long and hard over the exact nature of the associations between stone tools and animal bones with "cutmarks" found nearby. However, microscopic examination of the V-shaped grooves made on bones by stone tools readily discriminates this kind of damage from that made, for example, by the teeth of carnivores and scavengers. On the other hand, it is more difficult to tell whether an animal was dispatched by the hominids that butchered it, or whether the remains were simply scavenged. But given the small body size and the general vulnerability of hominids at this early time, the hunting alternative is generally considered rather implausible. Another factor is the prevalence of "torsional" fractures along the length of limb bones found at several early archaeological sites. Produced by pounding the bones with a rock, these fractures suggest that the hominids were in search of the bone marrow, a rich source of fats and proteins. This is one of the few resources represented by a dead animal that couldn't have been reached by most carnivores, and for which hominids would have been in competition only with bone-crunchers such as hyenas. Here is another factor suggesting that the hominids were part of the scavenging rather than the hunting brigade, at least as far as medium-sized and larger mammals were concerned.

Scavenging lion kills or natural deaths sounds like a rather humble and undemanding occupation, but actually the invention of cutting tools probably had a major impact on the lives of our earliest precursors. Principally, it made available a highly nutritious source of food, and allowed the hominids to dismember even quite large carcasses and to carry parts of them away to safe places for consumption. This was important, because the wooded-to-open regions in which many carcasses were presumably found were dangerous places for the small-bodied hominids to be, teeming as those places were with unfriendly animals with big teeth. We have no idea what proportion of the early hominids' diet was typically furnished by animal products, but we can surmise that the new way of life involved the adoption of new skills. These might well have included the interpretation of indirect signs in the environment as guides to the location of carcasses. Vultures wheeling in the sky overhead are a dead giveaway, as it were, and they might have furnished such an element. However, to go any further than this and to suggest, for example, that the early hominids were reading animal spoor, probably goes much too far in the direction of viewing these creatures as junior-league versions of ourselves. This is a major trap into which many an archaeologist has fallen.

So much for lifestyle. But what does the act of toolmaking itself tell us about the cognitive abilities of these early precursors? Quite a lot, as it turns out. Even otherwise gifted chimpanzees have proven to be failures as stone toolmakers (though some natural populations of these apes do use twigs to "fish" for termites, for instance, or use stones as anvils on which to break hard-shelled nuts). It turns out to take a fair bit of insight to figure out at precisely what angle to hit one rock with another to detach a suitable flake, let alone to identify what kinds of rock will make the best tools. Very early archaeological sites usually consist of little more than cutmarked animal bones and/or stone tools and the "debitage" (stone fragments) from their manufacture, just lying around on the landscape. In several cases, archaeologists have been able to piece together entire cobbles from the flakes produced by hominid activity at butchery sites; and often these rocks are foreign to the places where they were found. Indeed, in some instances, the nearest source of the rock is miles away. This is clear evidence of several things. First, that the hominids anticipated that they would need tools. Second, that they knew how to select suitable materials and were prepared to carry them for long distances before making them into tools as needed. And third, that the early toolmakers had the manual and cognitive skills necessary for this last activity. Whacking one cobble with another might not sound like a particularly sophisticated enterprise, but in fact there are a lot of quite fancy inputs into stone toolmaking, even of the simplest kind. And any stone toolmaker, particularly a spontaneous one, is very distinctly out of the league of all living apes studied so far.

How our early precursors came by these abilities is unfortunately still a matter for conjecture. It seems reasonable to conclude that these key achievements may not have sprung forth fully fledged, and that they may have been preceded by various simpler technological developments, perhaps such as we see among some chimpanzees today. But, quite frankly, we have no idea at all as to exactly what such developments might have been, though various options are currently being explored, mostly by way of both captive and naturalistic studies of apes.

Well, if we know rather little about the way in which the earliest stone toolmaking behaviors emerged, do we at least know *who* the makers were? Er, no, not really. In 1964 Louis Leakey and colleagues established the species *Homo habilis* for the more lightly built hominid fossils found in the lower levels at Olduvai Gorge; and ever since then the picture has been getting more complicated. The Olduvai fossils, a bit less than 2 myr old, were found alongside the crude Oldowan tools (the first ever identified as such) that occur at the bottom of the Gorge. And it was because of the presumed association between the two that the fossils were both admitted to *Homo* (following the then-voguish concept of "Man the Toolmaker"), and named *habilis.* There was initial resistance to this new species, the most ancient representative of our own genus; but as more fossils were found in the 2.5 to 1.8 myr interval, which happens precisely to match the age of the earliest tools from Africa, many were admitted into the embrace of *Homo habilis* and other species of "early *Homo.*" By now there is such a variegated assortment of specimens in this category that nobody really knows how many species are represented amongst them. However, they clearly include a rather large range in brain and tooth size and in bony anatomy, and the latest suggestion is to remove most or all of them from *Homo* and to place them into *Australopithecus* instead. I'm happy to go along with this, although I must note that, while tidying up the concept of *Homo,* this makes the notion of *Australopithecus* even messier than it was before (see the caption to the illustration on page 127). What's more, nothing changes the fact that there are no really good associations between any fossil hominids and early stone tools. Indeed, the closest thing to such an association comes from the Bouri area, in Ethiopia, where 2.5-myr-old cutmarked bones and stone tools have been found in fairly close proximity to some rather fragmentary fossils that were assigned to the new species *Australopithecus garhi.* The discoverers were quick to point out that there is a difference between proximity and association; but, as I suggest in a moment, it actually seems reasonable to conclude that, if we knew exactly who they were, we would want to classify the first toolmakers in *Australopithecus* rather than in *Homo.*

The fragmentary evidence of the body skeleton, for what it's worth, also supports the notion that the first toolmakers were of rather archaic build. A partial and very rubbly skeleton from lower Olduvai, attributed to *Homo habilis,* indicates that its owner had possessed extremely *Australopithecus*-like body proportions; and isolated limb bones from sites in Kenya dating from a bit less than 2 myr ago seem to indicate the same thing. What's more, there is nothing at any site over 2 myr old to suggest that any hominid of modern body structure was around when the earliest stone tools were made. Thus we have to conclude, at least for the time being, that stone-tool technology was the intellectual offspring of hominids with archaic body proportions, and who probably also possessed small brains. We should not be surprised by this, especially if we can get away from the "Man the

Toolmaker" mentality. After all—and this is something that is crucial to appreciate—any innovation, whether physical or technological, has to arise initially *within* a preexisting species—for where else can it do so?

The First Humans of Modern Body Form

For all the variety of the australopiths, then, these hominids maintained a pretty consistent body structure, one that evidently served them very well in their preferred forest-edge habitat and in other environments in which they may have found themselves. But an astonishing discovery made in the mid-1980s at a site in northern Kenya showed unequivocally that finally a new kind of hominid had arrived on the scene. For some time, crania had been known from the region around Lake Turkana, and from the period beginning about 1.9 myr ago, that nobody had any trouble classifying in *Homo.* The crania of the australopiths were small-brained, and had basically apelike proportions, with a big face hafted on to the front of a relatively diminutive braincase. The new Kenyan hominids, in contrast, were of much more humanlike proportion, with smaller faces and larger skull vaults. Brain size was edging up around 800 ml and beyond, not much more than half the modern size, but much bigger than the 350–500 ml of the australopiths. There is, once more, some variety among the cranial specimens, but everyone is comfortable calling them all *Homo.* All well and good, but there was still a limited amount that it was possible to say about these new hominids, particularly since they did not seem to be associated with any notable advances in stone-tool technology or butchery techniques. They continued to use essentially the same technical equipment that their prececessors had been wielding for more than half a million years, and it was difficult to infer that any substantial improvements in lifestyle had accompanied the acquisition of their larger brains.

All this changed dramatically with the unprecedented discovery, also in the Lake Turkana basin, of the skeleton of an adolescent hominid: the Turkana Boy, whom we've already met. This 1.6-myr-old skeleton is almost complete, missing, as I've noted, little more than the hands and feet. Not only are finds of this completeness vanishingly rare in the hominid record, but the skeleton itself presents a total contrast to anything known from earlier times. While earlier hominids had been tiny-bodied, standing only about three to four and a half feet tall, here was an individual who (though he died at only about nine years of age, when he stood about 5′3″) would have topped six feet if he had survived to maturity. What's more, a few details aside, this was an individual with an entirely modern body structure. In nearly all important features, his skeleton was substantially similar to our own.

And, just as interestingly, the Boy was built with long, slender limbs, just like the people who live in the hot, arid, and largely treeless climes around Lake Turkana today. This was no individual in transition between archaic australopith body form and our own; rather, it appears from this find that all of the essential modern features of body size and morphology had already shown up as an integrated package. Further, there is no obvious precedent for these new features in the fossil record.

Despite the apparent lack of any associated technological improvements, the Boy's body build by itself must be telling us something about lifestyle differences. Because here, for the first time, was a hominid who was clearly at home out on the hot, shadeless savanna, far from the sheltering forest. This was no simple refinement of the australopith theme; this was a different creature altogether. Built to shed heat, tall, and with long, striding limbs, the Turkana Boy was the first hominid we know of who must truly have been emancipated from the forest edges. And once this emancipation was achieved, humans were free to indulge their wanderlust; for, right after the emergence of this new kind of human, we begin to find evidence of human penetration out of Africa and into the far reaches of Asia.

Understanding this new phenomenon involves acknowledging that in the evolutionary process radical physical reorganization may sometimes occur as the result of relatively minor genetic changes, as we saw in an earlier essay. But unfortunately, although it's evident that the Boy represents more than just Nature's tinkering with a preexisting theme, the larger picture here is not as clear as we would like. For example, in his skull and tooth structure, the Boy himself is not exactly like the various 1.9–1.7-myr-old specimens from Turkana that also represent undoubted *Homo* (and which also vary considerably among themselves). Indeed, in many ways the Boy is quite distinct from them. It thus seems evident that, once the new *Homo* theme had emerged, Nature undertook its customary bout of speciation and experimentation. It's simply unfortunate for paleontologists, who would like to know more, that the Boy is the only individual of his general kind for which we have a clear association between the skull and the body skeleton. We may surmise that such famous crania as the unimaginatively named ER 3733 (and other specimens often regarded as belonging to the species *Homo ergaster*) would have possessed a body form broadly similar to the Boy's; but, regrettably, we can't yet be absolutely sure. All we can say is that, by about 1.6 myr ago, our precursors had somehow made an unprecedented leap in bodily structure.

Unhappily, for the time being we can't even make any intelligent guesses as to where this new structure came from. Although the Boy and other early unequivocal *Homo* must ultimately have had origins somewhere among the australopiths (the only reasonable candidates for their ancestry), at the moment we are totally unable to discern where those origins lay. Loath as I am to admit it, none of the australopiths known so far makes a particularly good candidate for the ancestry of *Homo* properly defined.

Patterns in Early Human Evolution

Looking at the first 3 myr of human evolution, then, we find that really meaningful change has been episodic indeed, and quite rare. There is no suggestion whatever in the record as it has come down to us of slow and progressive refinement, in one feature or in many. At the origin of any major group we expect to find a major innovation that set the path for future developments. In our case, this new introduction appears to have been the adoption of a habitual upright posture, though whether in the context of locomotion or something else remains equivocal. And this basic upright structure appears to have been very successful: It provided the platform for the production of numerous new species, while at the same time remaining essentially stable for a very long period. We don't know when this innovation first emerged, or anything about the morphology of the first organisms that possessed it, but it is reasonable to guess that the anatomical modifications involved came about as a result of what was a relatively minor change at the genetic level. What we do know, though, is that the innovations involved were established early on. Further, as far as we can tell, they were apparently not greatly improved upon over an extended period—during which Nature undertook little more than minor tinkering with other hominid body systems, despite producing a variety of species in different parts of the African continent.

The invention of stone tools was clearly an epochal event in human history; but it followed habitual upright posture by 2 myr or more, and has no clear antecedents in an admittedly imperfect record. Given that a whole variety of cognitive and manipulative innovations must have been involved in the inauguration of technology, it might seem surprising that, once the breakthrough had been made, little changed for the next million years. The ability to cut introduced previously undreamed-of possibilities, and clearly was exploited to good effect; but once the theme was established there is little evidence that it was elaborated or extended until a very long time had elapsed. Following the origin of stone tools about 2.5 myr ago, a million years were to pass before a new concept of tool emerged. This concept was radical, however. It involved not simply going for an attribute (a cutting edge) on a small flake whose exact shape was unimportant. Instead, it implicated the production of a larger, teardrop-shaped tool, fashioned symmetrically on both sides to a standard form. Even then, though, this new tool type persisted in its turn for a long period before itself being replaced. Thus the early history of technology is no more a story of gradual refinement than is the anatomical history of hominids: A good idea is apparently a good idea wherever it crops up, and if it is to be superseded, the new idea has to be a very good one indeed. Or perhaps, in human prehistory, good new ideas were simply few and far between. Whatever the case, the pattern that we observe in the record is the same: Truly significant innovation is both sporadic and rare.

The anatomical gulf between the australopiths and early uncontestable *Homo* such as the Turkana Boy yawns wide indeed. Yet we search in vain in the admittedly spotty record before the Boy for anything that we could interpret as his ancestor. Yes, the australopiths were bipeds of some kind, for there is little question that from the beginning these hominids were upright movers, with pelvic characters in particular indicating their commitment to an essentially orthal posture. But the earliest hominids had their own way of doing business, and it was distinctly different from ours. The Turkana Boy, in contrast, was very emphatically one of us: a creature clearly adapted for shedding heat and for life away from the forest. Once more, a major innovation in the hominid record had appeared abruptly, rather than as a result of continual refinement over the ages. As with the other major changes we've looked at, there is no convincing transformational signal here. It may seem somehow vaguely belittling to reduce the first half of hominid evolution to three basic changes: uprightness, toolmaking, and striding bipedalism. Yet, in their essentials, these were indeed the major innovations that counted. The appearance of our own kind apart, nothing else has impacted human history quite as strongly. In parallel, of course, the basic background tickover of new species, and of competition among them, certainly went on, and had an ongoing effect on how things turned out; but in the end it was this limited number of episodic but momentous changes that really made the difference in hominid evolution. On an individual level the tragic personal saga of the Turkana Boy is a moving reminder of the drama that marked the hard and short lives of our precursors—and it is also a reminder of how greatly we in the western world are nowadays insulated from the ravages of Nature. But at the same time the Boy serves as the best fossil exemplar we have of the dramatic way in which truly significant change tends to occur in evolution.

Chapter Five

The Enigmatic Neanderthals

Neanderthal. What more evocative word is there in the English language? Yet how many of us can specify exactly what it is that this magical name evokes? To some, the term Neanderthal is little more than a byword for benighted bestiality. At the same time, many scientists have devoted their careers to demonstrating that these extinct humans were simply a variation on the theme of our own *Homo sapiens.* Some view the Neanderthals as the ultimate victims, prey to the modern humans—the Cro-Magnons—who invaded their European and western Asian homeland some 40 kyr ago. Others see them as at least partly ancestral to the modern human inhabitants of Europe. Some see the archaeological record of the Neanderthals as testament to an intellectually lively, creative people; others, as the record of hominids who signally lacked the creative spark that distinguishes modern *Homo sapiens*. Whatever the case, the story of the Neanderthals is a dramatic one, the drama emphasized by the richness of the record they bequeathed us. So where does the truth about the Neanderthals lie? It would be foolish to claim that today we have anything close to a definitive answer to this question—although it begs to be asked, as I do here. But in attempting to answer it we have to bear in mind that the Neanderthals were not in any way an isolated phenomenon that came out of nowhere. Indeed, to understand these hominids involves delving into the past to a time well before any Neanderthals existed.

Human beings of essentially modern body structure are presumed to have become established in Africa at about 1.8 myr ago. But soon after this event, Africa, the birthplace of mankind and for well upwards of 2 myr its sole home, at least temporarily lost its centrality in the hominid record. This is partly for reasons of geological accident, for following about 1.4 myr ago there are simply fewer rocks of the kind likely to contain hominid fossils exposed on the African landscape. But partly it is also because hominids made their first exodus from their continent of origin at some time shortly after about 1.8 myr ago. This latter date is still a bit of a guess, but it does tie in pretty well with the time at which we presume essentially modern bodily anatomy emerged, and also with the steady trickle of early dates that has been coming in over the past few years for hominids in the Caucasus and points east. Thus it seems a pretty good bet that the initial human diaspora out of Africa was closely associated with a new lifestyle in which tall, striding hominids were freed from the wooded environments that had limited their ancestors. These new hominids, it appears, were able to move around pretty much as they pleased on the relatively open African savannas—and well beyond. Our well-known human wanderlust goes back an unexpectedly long way!

But while the former Africans evidently penetrated rapidly as far as Indonesia and the eastern reaches of China, evidence of their very early presence in Europe is more or less absent. Apart from claims of crude tools at a locality in France said to date from over 2 myr ago, and a couple of archaeological sites in France and Italy that might (or might not) be as old as 1 myr, there is nothing in the record to suggest an occupation by hominids earlier than about 800 kyr ago. Why this relatively late date is unknown, though it seems reasonable to guess that the environment may have been a big factor. The first African emigrants were, as far as we know, quite unsophisticated technologically, bearing only an Oldowan-level stone industry; and it thus seems likely that they did not not possess particularly efficient cultural means of coping with extremes in the external environment. However, while any early hominids who headed due north out of Africa would quite rapidly have encountered some pretty severe climates, those who turned right could have expanded their populations eastward while staying in the friendlier and more familiar subtropical zone. Whatever the case, we have to wait until a million years after hominids first quit Africa to find fossil evidence of the earliest Europeans. And even then, we don't know for sure whether these fragmentary 800-kyr-old specimens from Spain's Gran Dolina site are truly early Europeans, in the sense that they belonged to a population that was directly ancestral to the later occupants of the region. They might equally well have represented an initial attempt at colonizing the subcontinent that ultimately failed. Their morphology doesn't obviously anticipate that of later Europeans, and it was only about half a million years ago, 300 kyr after Gran Dolina times, that the archaeological and hominid fossil records began to become a regular feature of the European scene.

The exact affinities of the Gran Dolina hominids will almost certainly remain a puzzle for some time. But the picture subsequent to about 500 kyr ago is becoming at least a little clearer. When my colleague Jeffrey Schwartz and I started to look at the European hominid record a few years back, we knew that we should expect to encounter a substantial anatomical variety among the fossils comprising it. But, unlike hominid diversity in some other areas of the world, the variety we saw soon began to take on a fairly coherent form. The best-known of the ancient inhabitants of Europe are the Neanderthals. These hominids were highly distinctive, but with brains as large as our own. The now-extinct Neanderthals, members of the species *Homo neanderthalensis,* flourished widely in Europe and western Asia (at least as far east as the Altai Mountains in the northern Caucasus) from something over 200 kyr ago to about 27 kyr ago.

Paleoanthropologists had long recognized that certain earlier hominids from Europe, while distinct from the Neanderthals, also had features in common with them. In good linear tradition, however, they responded to this perception by giving them such names as "pre-Neanderthals" or "Protoneanderthals," implying that all these hominids had been part of the same steady progression. What Schwartz and I saw among the fossils, though, was something rather different. It was certainly not the trend toward full-blown Neanderthal status that might even, some suggested, have eventually culminated in ourselves. Instead, we concluded that several species of hominid were represented in the European record, all of them ultimately descended from the same common ancestor. The Neanderthals were simply the best-known and latest-surviving of the descendants of this ancestor.

This all makes excellent sense, because it's exactly what we would expect to happen when a new group of organisms colonizes a virgin territory, and begins to explore all of the new possibilities available to it. And although the details are rather fuzzy, what happened among the hominids in Europe was evidently the routine phenomenon of "adaptive radiation," which has been documented in one diverse group of organisms after another. The story is simple and straightforward: An ancestral hominid colonized Europe at about half a million years ago, and in this formerly *Homo*-free environment it ultimately produced a variety of descendant species that competed among themselves for precedence. We should bear in mind here that conditions at this time, during the regular 100-kyr cycling of the Ice Ages, were ideal for the production of new species. For the habitat constantly fluctuated, and human populations must have been regularly fragmented and recombined by factors entirely beyond their control—with all the consequences I noted in an earlier essay. And in the resulting ecological contest among the newly established hominids of Europe it seems that, in the end, one species, *Homo neanderthalensis,* won out over the others.

Nothing lasts forever, though, and the Neanderthals eventually found themselves battling for ecological space and survival against our own kind, *Homo sapiens.* This new species had evolved somewhere outside Europe, and it began expanding into the Neanderthals' heartland about 40 kyr ago. What's more, the invaders arrived not just with modern anatomy but with the whole panoply of behaviors that makes our species so remarkable today. *Homo sapiens* is without doubt not only a uniquely gifted but a uniquely dangerous creature; and the outcome of the resulting confrontation with the resident Neanderthals was nothing short of inevitable. Still, ourselves apart, *Homo neanderthalensis* is incomparably the best-documented hominid species that we know about, either behaviorally or anatomically. Also, as we've seen, Neanderthals were large-brained, and they inhabited the same region as some of our own ancestors at the same point in history. Interaction between the two species was thus inevitable; and it is in this extraordinary concatenation of circumstances that the peculiar fascination of the Neanderthals lies. History has made these now-extinct relatives unquestionably the best mirror that we possess to hold up to ourselves, and a singular resource to help us discover wherein our own remarkable uniquenesses lie. And maybe there's a touch of guilt tucked in there somewhere, too.

Who Were the Neanderthals?

Spotting species among fossil hominids is not always an easy task. All individuals of any species differ from one another in a host of minor ways, and sometimes it is tricky telling these kinds of differences from those that truly distinguish among species. This is especially true when available fossils are sparse and fragmentary, which is only too often the case. No such problem with the Neanderthals, however; and despite lingering claims to the contrary there can be no question that these hominids constituted their own distinctive species. Indeed, just as their distinctiveness as a group can hardly be questioned, the Neanderthals are so well known that we have a very good idea of how they tended to differ from one another in their bony anatomy. The upshot is that even those who would wish to demote the Neanderthals to the lowly status of a bizarre form of ourselves (the subspecies *Homo sapiens neanderthalensis*) have no trouble in recognizing a Neanderthal as a Neanderthal, and a modern as a modern, whenever they see one of either.

Among other things, while having large brains like us, the Neanderthals housed those organs in a long, comparatively low skull with bulging sides and that had a profile which retreated strongly behind the brows. The brows themselves are adorned with strong, rounded, bony ridges that form a double arch above the orbits. The face projects quite markedly in front of the braincase, but seen from above it is wedge-shaped, with a large, prominent nasal region (containing some unique structures), from which the cheekbones sweep back dramatically at the sides. The rear of the skull typically protrudes somewhat, too, and bears a curious small depression in the midline. In all of these characteristics, and in many more, the Neanderthals contrast with modern humans, who have relatively tall, short, and straight-sided crania with rounded posterior profiles and a small face tucked below the front end of the braincase. Our foreheads are vertical, and our browridges, if we have any to speak of, are divided into central and lateral portions. And so on, the general point being that, anatomically, the Neanderthals were distinctly different from us—and, just as importantly, from all other known hominids too, although they did share *some* of their distinctive features with their European relatives. The further implication of all this is that the Neanderthals would not have been reproductively compatible with the modern humans who arrived in their territory some 40 kyr ago—an inference with considerable impact on scenarios of what might have happened then.

Where the Neanderthals actually came from is unknown in detail. But both direct fossil evidence, and inferences from molecules recently extracted from the original Neanderthal specimen (discovered in Germany in 1856) and other Neanderthal bones, agree fairly well on the broad lines of Neanderthal origins. The lineage that led eventually to *Homo neanderthalensis* emerged somewhere in greater Europe around half a million years ago or a bit (possibly a lot) more. For the first part of their tenure in Europe the Neanderthal ancestors had to contend with related species as they explored the environment they had inherited; but by something like over 200 kyr ago, when the first typical Neanderthal fossils show up in the record, the competition seems to have disappeared, and all subsequent indigenous European fossils known fit quite comfortably into *Homo neanderthalensis.* It's often claimed that these denizens of northern latitudes were "cold-adapted," comfortable in the severe climates they sometimes inhabited; but the fact is that it's not clear in exactly what kind of environmental circumstances the Neanderthals originated. By Neanderthal times the European climate was, of course, locked into the cycle of the Ice Ages, whereby the polar icecaps expanded on an approximately 100-kyr rhythm, taking worldwide temperatures down with them before those temperatures rose again and shrank the ice sheets once more. But these climate changes were not monolithic, and showed constant minor and not-so-minor oscillations. We would have to know with great precision exactly where and when the Neanderthals originated, to locate that event in this cycle—and, of course, we don't. What we *can* say, however, is that creatures readily recognizable as Neanderthals flourished not only through the coldest parts of the glacial cycle, but through warmer episodes that resembled the one we are enjoying today. The jury must thus remain out on whether cold adaptation was part of the Neanderthal armamentarium.

Neanderthal Technology

The first European hominids possessed rather primitive stoneworking technology. Indeed, the earliest stone tools from the subcontinent were hardly any more impressive than those that distant precursors had been making in Africa 2 myr earlier. But at some time between about 300 and 200 kyr ago, a major innovation crept into the European record, possibly imported from elsewhere. This was the "prepared-core" tool. In Africa Oldowan-style technology had been joined about 1.5 myr ago by the wildly successful Acheulean hand axe and its derivatives, a category of stone tool whereby a substantial cobble, possibly derived itself from a bigger rock, had been carefully shaped on both sides to a standard tear-shaped format. Although the name Acheulean derives from a site in France, such tools actually took a very long time to catch on in Europe, and barely did at all in parts of Asia. Not long after the arrival of the Acheulean in Europe, however, we find prepared-core tools turning up in the record; and even if this new type of tool was not actually a

Neanderthal invention, it quickly became a Neanderthal hallmark as part of their Mousterian stoneworking industry, named for a site in western France.

The prepared-core tool marked a major leap not only in technology but in cognitive capacity. For if Oldowan toolmakers had simply been going for a sharp flake, irrespective of what it looked like, while Acheuleans patiently bashed away at a core until the desired shape emerged, this new type of tool represented an entirely new concept. A stone core was exactingly shaped until a single blow could detach a flake that was of predetermined size and shape. Each such flake represented a nearly finished tool. Implements of this kind were razor-sharp all around their periphery, and the flakes could easily be trimmed to the precise final product wished for. If made out of particularly desirable materals (flint or chert, for instance), such tools were regularly resharpened, increasing the apparent variety of form—and giving archaeologists a misleading notion about how many specific kinds of tools made up the Mousterian tradition, which had become the standard in Europe by about 200 kyr ago.

Stone tools are, obviously, a pretty indirect reflection of the ways of life in which they were used, though usually they are a substantial part of what we have to go on. Still, they repay close examination. Thus studies of the wear on stone tools from Mousterian sites suggest that they were used for scraping skins and shaping wood, among other purposes. Wood itself must have been an important resource for Neanderthals, but unfortunately its preservation over the millennia is vanishingly rare. However, a recent astonishing find at a 400-kyr-old (hence pre-Neanderthal) site in Germany may reveal what kind of woodworking might have been available to Neanderthals. At the locality of Schoeningen, some miraculously preserved wooden spears have been found. Some are well over six feet long, with carefully fire-hardened tips, and are shaped so that the center of gravity is forward, as in a modern javelin. These were spears for throwing, not for thrusting, and they imply that their makers possessed quite refined hunting techniques—more sophisticated at this great antiquity than we might otherwise have guessed.

The period around 400 kyr ago also contains some other novelties. It has been claimed that, at about this time, rudimentary huts were built at Terra Amata in southern France. These structures possibly even contained hearths where fires burned. There also seems to be supporting evidence for such activities only slightly later in time, from the 350-kyr-old German site of Bilzingsleben. Within the Neanderthal period itself, there appears to be rare if increasing evidence that these extinct hominids rigged up presumably hide-covered structures at their occupation sites. Interestingly, though, the structures of those occupation places during Neanderthal times tended to be rather haphazard: Tools, bones, and other detritus were rather randomly scattered around the living space. The more compulsive organization that we typically see at *Homo sapiens* living sites is rare among Neanderthals, though at the Israeli Neanderthal site of Kebara it is clear that fires burned in one area, stones were knapped somewhere else, and animal bones were thrown toward the back wall of the cave.

The Neanderthal Way of Life

How expert the Neanderthals were as hunters has been widely debated, though it certainly seems that in all their intricacies the sophisticated hunting techniques of our own species are the unique property of *Homo sapiens.* Still, this doesn't mean that all previous hominid species were clods in this department. Indeed, the Mousterian evidence suggests that the Neanderthals varied their economic activities from time to time and place to place, just as you might expect them to in an extended period of oscillating climates. And, given the vast and topographically varied area they inhabited, from the Levant in the south to Wales in the north, and from the Altai all the way west to the Atlantic, it is hardly surprising that it is tough to identify a "typical" Neanderthal lifestyle. If you really want generalizations, though, it does seem likely that, in contrast to modern hunters and gatherers, who are (or were) "collectors," carefully monitoring the resources around them, the Neanderthals were "foragers," simply exploiting what they came across in their daily travels. Yet even this may be an oversimplification, and we really have very little idea exactly how the Neanderthals typically accumulated the animal bones that littered their living spaces. In a single area in Italy, for example, there is evidence for Neanderthal occupation at about 120 kyr and 50 kyr ago. Archaeologists believe that in the earlier time there is better evidence that the carcasses were scavenged, while the later ones were probably hunted. On the basis of the nature of the respective bone litters this certainly seems a reasonable conclusion to draw; but it's also true that, since Schoeningen times at least, we have presumptive evidence for a potential ambush-hunting ability. Thus the jury will, alas, have to stay out a little longer on the matter of Neanderthal subsistence patterns and their possible improvement over time. All we can reasonably surmise is that, while it is the bony evidence associated with meat that best survives—and some recent isotopic analyses suggest a quite highly carnivorous diet—vegetable products almost certainly contributed significantly to the Neanderthal diet in most times and places as they have done, historically, among all but the most specialized of modern human hunting and gathering societies. Such considerations apart, however, it also seems reasonable to conclude that, whatever Neanderthal lifestyles were, they were foreign to our own. This is not to imply that they were ineffective. After all, the Neanderthals were for a long time hugely successful, over a vast and complicated area of the Earth's surface; and ours is certainly not the only way of doing business in this world.

The Neanderthal Way of Death

One reason we know so much about the Neanderthals is that sometimes they buried their dead. Earlier Neanderthals are considerably less well-known than later ones, and are generally found in different sorts of contexts; and this difference is probably due to a later trend toward burial, which had the effect of enhancing the survival of skeletons over the long term. Nobody knows exactly when burial was invented (or by whom; some of the earliest burials come from the Levant, where they did not involve Neanderthals); but it is certainly within the Neanderthals' tenure on Earth that inhumation became at least occasionally practiced. Once introduced, it may have become somewhat more common as time passed.

What burial *meant* to the Neanderthals is, of course, another story. To us, this ritual, almost universal in one form or another among modern humans, is loaded with overtones of grief, loss, and, as often as not, with implications of transition, of an afterlife. If you're capable of spiritual feelings, there is no more efficient mechanism than a funeral to bring them out. But burial doesn't necessarily have to do with any of this. Remember that Neanderthals often found shelter in the kinds of places (cave mouths, especially) that were also favored by carnivores and scavengers. Perhaps burial was one means of discouraging the attentions of such creatures. Or maybe it was simply a pragmatic way of relieving valued living spaces of a rather distressing kind of clutter. Or again, maybe it was a way of dealing with obscure emotions, of expressing grief and loss. And none of these motivations need necessarily have been connected to feelings that we would recognize as spiritual awareness.

One reason for believing that they weren't, indeed, is implicit in the nature of the burials themselves. Early modern humans almost everywhere included "grave goods" in burials along with the corpse: decorative symbolic artifacts, or more humble objects from daily life, that would have been desirable to the deceased in an afterlife. And Neanderthal burials contain nothing of this kind, except for the occasional stone tool or fragment of bone that would anyway have been lying around on the cave floor, ripe to be kicked in by accident as the grave was filled. There are a couple of Mousterian burials from the Levant, both around 100 kyr old, in which the deceased appears to have been buried clutching an animal bone; but the individuals concerned were not Neanderthals, and in any case it's hard to tell what the significance of the bones was. Certainly, no overtly symbolic objects were involved. For this, we have to await the arrival of the behaviorally modern *Homo sapiens* who replaced the Neanderthals in Europe. These early people led lives, and underwent deaths, that were loaded with symbolism, and they were not reluctant to express this in their funerary practices. The most celebrated example of an elaborate early human burial comes from Sungir, in Russia, where almost 30 kyr ago an adult male and two juveniles were carefully interred with an astonishing variety of decorative and useful objects. The adult male, for instance, had been wearing a tunic on to which thousands of drilled and polished mammoth-tusk beads had been sewn.

Perhaps more suggestive than any Neanderthal burial, however, is the report from Iraq's Shanidar cave of the skeleton of an aged male who had suffered, perhaps from birth, with a withered arm. In the harsh world of the Neanderthals it is doubtful that so severely handicapped an individual could have survived without the consistent support of his group; and this in itself implies some degree of social caring, at least for the still-living. Shanidar is also the site of the famous "flower-burial," in which a male was interred in a grave that turned out to be rich in the pollen of spring flowers. There are numerous ways, of course, in which the pollen could have found its way into the grave; but if it was there because the individual had been laid to rest on a bed of flowers, the pollen would be as close to grave goods as any Neanderthal burial has come.

The Cro-Magnons

These, then, were the Neanderthals: a successful, large-brained group of early humans whose lifeways and technologies were at least as sophisticated as anything that ever went before them. Indeed, these human relatives were altogether admirable in many ways; and they had sailed without apparent difficulty through some extremely difficult environmental times. Yet what a contrast with the invading *Homo sapiens*—popularly called Cro-Magnons—who so rudely disrupted their tranquillity at some time after about 40 kyr ago! Among the Neanderthals, for instance, there is really no convincing suggestion of symbolic behaviors in the copious record of themselves they left behind. Starting with Bilzingsleben, some 350 kyr ago, an occasional plaque with odd scratchings or some polishing on it has been claimed as symbolic. But in the unlikely event that these are intentional symbolic objects at all, what they are remarkable for is their crudeness, and even more for the fact that they are very much the exception, rather than the rule. As far as we can tell from the preserved record, symbolism is highly unlikely to have been a routine or important factor in the Neanderthals' existences. The Cro-Magnons, on the other hand, led lives that were drenched in symbol. Nobody knows exactly where these new people came from, although the betting is that their ultimate place of origin was in Africa. But by the time they penetrated Europe they had

---

The art on the opposite page is one interpretation of relationships among the various fossil hominid species known to us (time is on the vertical axis, most recent at the top). The diagram is far from definitive, but it serves to provide a visual idea of the diversity of fossil hominids over time. It also includes a couple of developments that have occurred recently: the addition to the hominid roster of the new six-million-year-old species Orrorin tugenensis *(Millennium Man), substantially the oldest claimant so far to hominid status, and the new 3.5-myr-old genus and species* Kenyanthropus platyops. *Paleoanthropology has yet to digest the significance of either* Orrorin, *which its finders claim side-lines* Ardipithecus *and the australopiths from the main hominid stem, or of* Kenyanthropus, *which has been said to do the same thing. It's worth suggesting, however, that, curmudgeonly claims of "muddying the waters" to the contrary,* Kenyanthropus *may actually help to clarify the rapidly complexifying picture. This is because it may provide some additional context for the* Homo rudolfensis *skull ER 1470, which previously appeared as something of an anomaly. In any event, the positions of these new fossils in this diagram are highly provisional and dictated largely by time and intuition. Illustration by Bridget Thomas.*

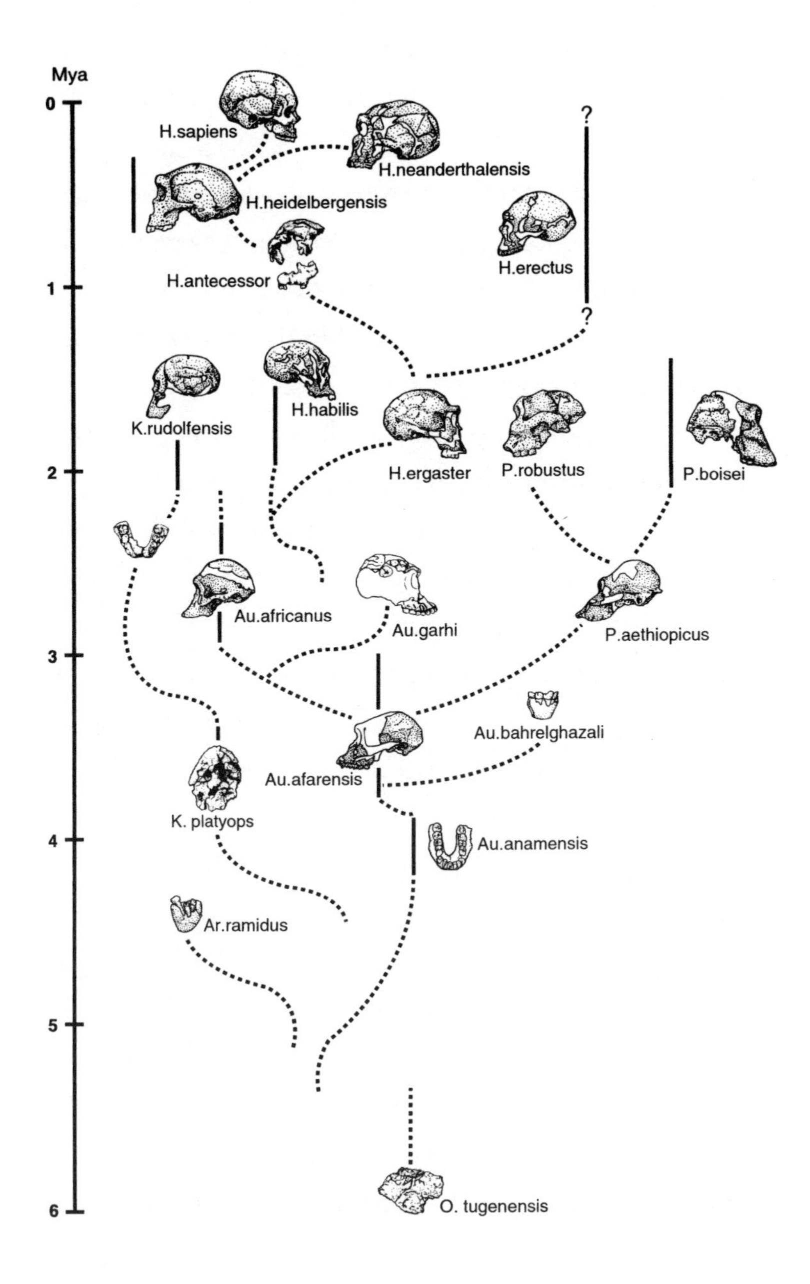
Mya
0
1
2
3
4
5
6
H.sapiens
H.neanderthalensis
H.heidelbergensis
H.erectus
?
?
H.antecessor
K.rudolfensis
H.habilis
H.ergaster
P.robustus
P.boisei
Au.africanus
Au.garhi
P.aethiopicus
Au.bahrelghazali
Au.afarensis
K. platyops
Au.anamensis
Ar.ramidus
O. tugenensis

acquired a fully modern sensibility, though it would be a big mistake to think that their beliefs necessarily resembled anything in our own twenty-first-century view of the world.

Unlike any hominids we know of who had preceded them, before 30 kyr ago the Cro-Magnons were making exquisite sculptures, employing both astonishing powers of naturalistic observation, and arbitrary stylization. They painted spectacular multicolored art on the walls of caves, with equal perspicacity. They made music using sophisticated wind instruments, and probably percussion as well. They engraved notations on bone plaques that clearly involved some form of record-keeping. And, as we've seen, the Cro-Magnons often buried their dead quite elaborately. They ornamented their bodies (dead or alive) with bracelets, necklaces, pendants, and a host of other decorative objects, and they also decorated everyday objects such as scraper handles. On the technological front, where Neanderthals had stuck to stone as a material for tools (though they did use bone or antler hammers in this pursuit), t he Cro-Magnons delighted in fashioning implements from soft materials, whose distinctive properties they obviously understood intimately. The Cro-Magnons soon thereafter invented bone needles, announcing the arrival of couture; and they even introduced kiln-baked ceramics. Further, around this time bird and fish bones begin to turn up at archaeological sites in some profusion. These are presumptive evidence of fancier subsistence activities, which are also revealed in some gigantic slaughter-sites of medium- and large-bodied mammals. And even if Neanderthals had caught fish, for example (bears do, after all), the evidence implies that they ate them where they were caught, rather than bringing them back to a central place for sharing, according to typical human hunter-gatherer practice. What's more, nobody will dispute that the Cro-Magnons must have possessed language more or less as we are familiar with it, whereas it's anyone's guess how the Neanderthals communicated. The list of unprecedented Cro-Magnon achievements could go on and on. But the point should already be clear: The Cro-Magnons were *us* in the most profound of senses, and they were the first kind of human we can hope to understand in terms of our own psychology and cognitive apparatus.

Neanderthal and Cro-Magnon Interactions

This was the new entity that first confronted the Neanderthals at some time around 40 kyr ago or a little less. What a shock this must have been for the latter! We have no idea what the Neanderthals were like temperamentally: whether they were aggressive or retiring; cooperative or individualistic; forthright or sneaky; trusting or suspicious; crude or lovable; or, like our own species, all of the above. For all that we know about the Neanderthals from the impressive record they left behind, we are still unable to form a psychological profile of these hominids that might help us to divine how they might have interacted with this new element on the landscape. When the tall strangers started to trickle into their formerly tranquil valleys, was the Neanderthals' response to hide from them, or to attack? To welcome or to reject them? To mingle with them or to run? To learn from grim experience and then run? To ignore them as long as they could? It is really impossible to know. In fact, we have only one source of evidence of the Neanderthal–*Homo sapiens* interaction, and it is one, alas, that is very difficult to interpret. At a few sites in France and elsewhere, there are industries that combine elements of both the Cro-Magnons' Upper Paleolithic culture and the Mousterian of the Neanderthals. The best known of these is the "Châtelperronian" industry of France, represented at a few rare sites in the 36–32-kyr range. As we've seen, the Neanderthals typically made stone tools by detaching flakes from roundish cores. The Cro-Magnons, on the other hand, used cylindrical stone cores from which they could detach numerous long, thin "blades." They also used bone and antler as materials, and made decorative objects. In the Châtelperronian about 50 percent of the stone tools were blades, which raised an obvious question about the industry's makers. And to complicate things still further, at one Châtelperronian site a few decorative objects were found. The question was inevitably raised as to whether the Châtelperronian might in some sense have been "transitional" between Mousterian and Upper Paleolithic, and its authors by extension "transitional" from Neanderthal to modern. But eventually it became generally agreed that the Châtelperronian was the work of Neanderthals, pure and simple.

Unfortunately, though, this didn't resolve the main question at issue: whether the Châtelperronian was in some sense "transitional" between the Mousterian and the Upper Paleolithic. These sites date from a time when we know that Cro-Magnons were already present on the landscape of Europe. And this is where the question of interaction enters in, for some form of interaction there must have been. The possibilities are numerous. Neanderthals were exquisite stoneworkers, and might have picked up blademaking and the use of bone directly from the newcomers; or they might have figured out how to do it from detritus left behind at Cro-Magnon campsites. Then again, some Neanderthal might have acquired the few decorative objects by trade (and remember, we're only talking of one site). Or maybe these objects were stolen, or were lost and found, or were forcibly taken in a violent interaction. Or maybe they were actually made by a creative Neanderthal individual. Unfortunately, we just don't know. All we can say for sure is that, whatever it was, the interaction did not last long, for the Châtelperronian made only an ephemeral appearance on the scene. What's more, the likelihood that any kind of long-term constructive interaction took place between the two hominid species is negated by the consistent pattern at archaeological sites, whereby the Châtelperronian, or more commonly the Mousterian, is abruptly replaced by the Upper Paleolithic. There is no evidence of intergradation or transition between the two, apart from the Châtelperronian itself. There are a couple of sites at which Châtelperronian and Upper Paleolithic briefly alternate; but again, the alternation is abrupt.

All this fits well with the only other thing we can be reasonably confident about in this hazy situation, which is that the Neanderthals and moderns did not interbreed. They are simply too different anatomically not to have belonged to different species. This is not to say that the occasional act of mating might not have occurred, but the two hominids are so distinct as to suggest very strongly that no biologically significant exchange of genes would have or could have occurred. Exactly where the barriers to mating might have lain is entirely conjectural, but there is little doubt that they were there. Still, any plausible alternative interactions to intermingling do tend to place *Homo sapiens* in a rather unfavorable light, and there are those who would prefer to think that the Neanderthals were "genetically swamped" by the invading moderns, rather than eliminated in more direct ways. This is why paleoanthropologists of linear mindset are fond of pointing to so-called Neanderthal characteristics in this or that early *Homo sapiens* specimen, or to "advanced" characters in one Neanderthal or another. But none of these instances is actually such as to suggest biological intermediacy.

The latest wrinkle in this argument is the announcement of an incomplete child's skeleton from Portugal that is dated at about 24 kyr ago and is alleged to show both Neanderthal and modern features. Those who favor this notion deal with the fact that the infant lived several thousand years after the disappearance of the last Neanderthals we know of by dismissing the idea that this was a "love child" between a Neanderthal and a modern, and claiming instead that it represents a hybrid population that had been quietly (and very discreetly) intermixing for four thousand years (two hundred generations, at least). Genetically, this is a rather bizarre claim, since after so many generations of interbreeding we would not expect to be able to identify Neanderthal versus modern characteristics in the descendant population. But special pleading is not unknown in paleoanthropology, though few are convinced that this infant is anything else than a chunky modern child: a descendant, indeed, of the very people who had shown the Neanderthals the door. The main scientific importance of the specimen thus lies in showing us once again how reluctant scientists can be to abandon cherished world views.

The Last Neanderthals

Judging by the frequency of known fossils, the period leading up to the coldest part of the last glaciation was a successful time for *Homo neanderthalensis.* More of these hominids are dated to between 50 and 40 kyr ago than to any other comparable interval—though this might in part be an artifact of the technique of radiocarbon dating, which reaches the outside of its useful range just before the Neanderthals start to disappear. Nevertheless, there can be little doubt that the 50–40-kyr period was the heyday of the "classic" Neanderthals of western Europe. Cold it might have been, but the Neanderthals themselves were flourishing. In retrospect, though, we can see this idyllic period as the the calm before the storm. The first dates for the Upper Paleolithic in both eastern and western Europe (Bulgaria and Spain, respectively) come in at about 40 kyr ago, although such early dates are rare, and we have no clear idea of where the invaders came from: whether first from the east, or from the southwest via Gibraltar, or even in the equivalent of a pincer movement. However, although Cro-Magnon sites do not become common in Europe until some time after about 35 kyr ago, once the newcomers became established they spread like wildfire, and Neanderthal localities began to dry up. One site after another shows many millennia of at least sporadic Mousterian occupation yielding abruptly to the Upper Paleolithic. By about 30 kyr ago Mousterian sites had become vanishingly rare, and mostly confined to the remote fastnesses of the Iberian peninsula. The latest dates we have for Neanderthal occurrence come from Portugal, notably at the seashore caves of Figueira Brava and Salemas (about 28 kyr ago), and from the rugged landscape of southern Spain. A similar date has recently come in for a site in Croatia. At the Andalucian cave of Zafarraya a few Neanderthals may have lingered (without any indications of cultural or biological exchange with Cro-Magnons) as late as about 27 kyr ago. But then, whether with a bang or a whimper, the Neanderthals were gone, forever.

Paleoanthropologists are, by and large, a pretty liberal and warm-hearted lot who don't like to dwell too much on the darker side of our species. And it is certainly much nicer to contemplate the supposed genetic swamping as a putative reason for the disappearance of Neanderthal morphology from the record than it is to consider any of the alternatives. Regrettably, though, it is the alternatives that we have to confront, given both the lack of any evidence for a biological transition from Neanderthals to moderns, and the inherent improbability of this event. So, what might those alternatives have been? The least offensive of them is that *Homo sapiens* was simply a more efficient exploiter of the environment than *Homo neanderthalensis* was, and gradually—and thoughtlessly—edged the latter out, with a minimum of direct interaction between the species. Well, it might have happened that way; as indeed it might well have been that the Neanderthals had evicted those who had occupied the European continent before them. But in the whole imponderable story of the demise of the Neanderthals themselves, there is, unfortunately, one thing we know only too well. And that is the appalling historical record of *Homo sapiens.* Human beings move around with astonishing mobility; and groups of modern human invaders have compiled an almost unrelievedly abominable archive in their behavior toward resident human populations, let alone toward other species. Think of the horrendous savagery of the Vikings, or the Mongols, or the Crusaders who sacked Jerusalem, toward those whose territory they invaded. Often such excesses are justified by denying the residents "humanity"—and this, of course, would have been even more easily done in the case of Neanderthals than in that of *Homo sapiens* belonging to unfamiliar cultures. The Cro-Magnons were admirable in many ways, as indeed are (almost) all *Homo sapiens.* But we shouldn't be misled by the ethereal art of Altamira, Lascaux, and Chauvet into embracing the notion that the Cro-Magnons were incapable of behaving every bit as abysmally as humans are known to have in historical times (think Rwanda and Bosnia).

How the Neanderthals might have reacted to this new human phenomenon we simply don't know. They may have defended their territory doughtily. But in the end they evidently didn't stand a chance against the craft and guile of the invaders, with their linguistic and symbolic skills. Europe is a large place, with numerous obscure corners, environments, and topographic crannies; yet the fact is that the wholesale replacement of the Neanderthals did not take a hugely long time. And this in itself implies a direct rather than an indirect confrontation between the two kinds of hominid, neither of which is likely in those early times to have been very common on the landscape. Quite simply, *Homo sapiens* appears to be inherently—and probably inherently savagely—intolerant of competition from its relatives, as the Neanderthals so eloquently attest, and as our closest surviving relatives, the great apes, are still discovering in their turn.

Chapter Six

How Did We Achieve Humanity?

When we contemplate the extraordinary abilities and accomplishments of *Homo sapiens,* it is certainly hard to avoid a first impression that there must somehow have been an element of inevitability in the process by which we came to be what we are. The product, it's easy to conclude, is so magnificent that it *must* stand as the ultimate expression of a lengthy and gradual process of amelioration and enhancement. How could we have got this way by accident? If we arrived at our exalted state through evolution, then evolution must have worked long and hard at burnishing and improving the breed, must it not? Yet that seems not to be how evolution works; for natural selection is not—it cannot be—in itself a creative process. Natural selection can only work to promote or eliminate novelties that are presented to it by the random genetic changes (influenced, of course, by what was there before) that lie behind all biological innovations. Evolution is best described as opportunistic, simply exploiting or rejecting possibilities as and when they arise; and, in turn, the same possibility may be favorable or unfavorable, depending on environmental circumstances (in the broadest definition) at any given moment. There is nothing inherently directional or inevitable about this process, which can smartly reverse itself any time the fickle environment changes.

Indeed, as we'll see a little later, perhaps the most important lesson we can learn from what we know of our own origins involves the significance of what has in recent years increasingly been termed "exaptation." This is a useful name for characteristics that arise in one context before being exploited in another, or for the process by which such novelties are adopted in populations. The classic example of exaptation becoming adaptation is birds' feathers, as we have already seen. These structures are essential nowadays to bird flight, but for millions of years before flight came along they were apparently used simply as insulators (and maybe for nothing much at all before that). For a long time, then, feathers were highly useful adaptations for maintaining body temperatures. As adjuncts to flight, on the other hand, they were simply exaptations until, much later, they began to assume an adaptive role in this new function, too. There are many other similar examples, enough that we can't ignore the possibility that maybe our vaunted cognitive capacities originated rather as feathers did: as a very much humbler feature than they became, perhaps only marginally useful, or even as a byproduct of something else.

Let's look at this possibility a little more closely by starting at the beginning. When the first Cro-Magnons arrived in Europe some 40 kyr ago, they evidently brought with them more or less the entire panoply of behaviors that distinguishes modern humans from every other species that has ever existed. Sculpture, engraving, painting, body ornamentation, music, notation, subtle understanding of diverse materials, elaborate burial of the dead, painstaking decoration of utilitarian objects—all these and more were an integral part of the day-to-day experience of early *Homo sapiens,* and all are dramatically documented at European sites more than 30 kyr old. What these behavioral accomplishments most clearly have in common is that all were evidently underwritten by the acquisition of symbolic cognitive processes. There can be little doubt that it was this generalized acquisition, rather than the invention of any one of the specific behaviors I've just listed—or any other—that lay behind the introduction of "modern" behavior patterns into our lineage's repertoire. This new capacity, what's more, stands in the starkest possible contrast to the more modest achievements of the Neanderthals whom the Cro-Magnons so rapidly displaced from their homeland in Europe and western Asia. Indeed, Cro-Magnon behaviors—just like our own—evidently differed totally from those of any other kind of human that had ever previously existed. It is no denigration at all of the Neanderthals and of other now-extinct human species—whose attainments were entirely admirable in their own ways—to say that, with the arrival on Earth of symbol-centered, behaviorally modern *Homo sapiens,* an entirely new order of being had materialized on the scene. And explaining just how this extraordinary new phenomenon came about is at the same time both the most intriguing question, and the most baffling one, in all of biology.

Inevitably, of course, there are hints in the archaeological record well before Cro-Magnon times of behaviors in which we might perceive some degree of "modernity." The beautifully shaped 400-kyr-old wooden javelins miraculously preserved at Schoeningen in northern Germany, for example, have impressed some observers with their implication of remarkably advanced hunting techniques at a time when stoneworking methods in Europe still remained quite rudimentary. But their relationship to symbolic cognitive processes is more questionable: Intuitive mental operations can take one a very long way, and it's entirely plausible that these spears could have been invented, made, and used without the aid of symbolic cognitive functions. Slightly more recent in time, some scratches on bone plaques from another German site, Bilzingsleben, have been interpreted as symbolic markings—although there are other potential explanations for these incisions, and most feel that deliberate symbolism is a fairly dubious reading. However, even in the unlikely event that these markings were intentionally made, their crudeness and rarity do no more than emphasize that, at this early stage, any form of symbolic expression was very much the exception in hominid behavior, rather than the rule.

To put the matter in a nutshell, Cro-Magnon society was organized around a wealth of different kinds of symbolic expression, whereas in earlier times any putative examples of symbolic activities were isolated individual instances at best. Even the famous Neanderthal burials, or the occasional claimed instances of cannibalism by these hominids, only very arguably represented symbolic practices. What's more, the remarkable Cro-Magnon sensibility was quite evidently imported into Europe fully formed, which obviously makes the subcontinent not the best place in which to seek its antecedents. Instead, we should almost certainly look to other areas of the world for potential precursors of this new phenomenon. And this makes it particularly regrettable that at present the archaeological record which might potentially carry the signal of such events outside Europe is thin indeed. Of course, absence of evidence is not at all the same thing as evidence of absence, and the sparseness of the record is almost certainly due in large part to the vastness of the region involved and to the relative shortage of archaeologists working there. And there are nonetheless some suggestive observations that are well worth heeding.

The best such record, as far as it goes, is that of Africa. Technologically, one of the hallmarks of the Cro-Magnons as they entered Europe some 40 kyr ago was the manufacture of blade tools: stone flakes that were more than twice as long as wide, and were struck in multiples from cylindrical cores. And blades have now been reported from a site in Kenya that may be as much as 250 kyr old. As in the case of the Schoeningen spears, the relationship of such manufacture to the precise cognitive abilities of the toolmakers is less than entirely clear; but at the very least we can look upon blade toolmaking as some kind of a straw in the evolutionary wind. However, this is only part of the story. More suggestively, barbed bone harpoon points that may be 60–80 kyr old have recently been recovered at a site in the eastern Congo. These elaborate implements, made in a difficult material, look very similar indeed to points that began to be manufactured in Europe only about 20 kyr ago. In the same vein, there is evidence for flint-mining and the long-distance transport of exotic materials in Africa long before anything similar is observed in Europe. Hafted projectile points also show up quite early in Africa. Yet more compelling are deliberately incised ostrich-eggshell beads from a site in Kenya that dates from about 50 kyr ago, and pierced snail-shell disks, clearly intended for stringing, from an even older site in Ethiopia. And there is evidence, which will hopefully be more fully explored in coming years, of local differentiation of technological—and by extension cultural—traditions in Middle Stone Age Africa that greatly exceeds anything seen in northern continents before the Cro-Magnons. None of these African developments in isolation does much more than hint that this continent had once again played a central role in the emergence of new human behavior patterns; but, considered together, they provide a strong suggestion that this may have been the case. The exact pattern of acquisition of these various modern behaviors remains obscure, as does the process by which premodern populations in Africa became modern. The record, such as it is, certainly suggests that a variety of innovations were being exploited in Africa 200–100 kyr ago by a presumably equal variety of different human populations. But exactly what this evidently complex process consisted of, and how the various elements involved finally came together in the unique modern sensibility, remains an open question.

But Africa is not the only part of the world with rare and intriguing hints of early "advanced" behaviors. In Israel blade tools show up quite early, and more than 50 kyr ago, someone at the Levantine site of Quneitra deliberately carved a series of concentric markings on a flint plaquette. The stoneworking industry at this site is Middle Paleolithic, roughly equivalent to what Neanderthals were producing in Europe at the same time; the problem is, though, that we cannot be sure who engraved the Quneitra plaque, since both *Homo neanderthalensis* and *Homo sapiens* were around in the Levant 50 kyr ago, and the products of both, as far as can be told, were functionally identical. The same applies to the potential makers of pierced marine shells found in a Levantine burial that is close to 100 kyr old. Elsewhere, evidence of early symbolic production is even sparser. In Australia, for instance, some rock engravings that were thought by some to have been very early have now been redated to within the period, beginning some 60–50 kyr ago, when we know the continent was already inhabited by humans. What remains particularly intriguing about the early occupation of Australia, however, is that the immigrants must have crossed a substantial water barrier to get there. Even at times of lowered sea levels, Australasia has always been separated from the landmass to its north by at least sixty miles of open ocean, and any hominid crossing this stretch of water must have had formidable navigational skills. Few would contest that hominids capable of making this journey must have possessed effectively modern cognitive abilities, in which case hominid occupation of Australia by 60 kyr ago (or not much less) is fairly firm evidence for an early emergence of the human capacity. Still, while highly suggestive, this doesn't give us a great deal that's very specific to go on.

Neither, unfortunately, does the physical record. Fossils of hominids that existed at the time when we can infer that the modern human capacity was being formed are fairly thin on the ground, and their analysis has been handicapped by an excessively simplistic view of morphology. Thus paleoanthropologists have long been mesmerized by such features as "large browridges" (which modern humans are not supposed to have), and "chins" (which they are), without realizing that these terms in fact cover a multitude of different morphologies. In reality, modern humans may have big browridges or small ones; the important point is that all share the same very unusual brow structure, with a division of the ridges, whatever their size, into distinct central and lateral portions. Similarly, the human chin is not simply a bulge in the profile of the lower jaw (which many hominids have to a limited extent) but is a unique and complex structure that assumes the form of an inverted T. If we take such considerations into account, we discover that almost certainly a lot more was going on at and around the birth of *Homo sapiens* than has yet met the eye. Nonetheless, it is fairly clear by now that the earliest evidence of anatomically modern humans comes from Africa, or at least from the geographically adjacent area of the Levant, where modern human morphology was established in some populations at least by about 100 kyr ago. The conclusion of an African origin for *Homo sapiens* is also supported by DNA studies, which show that human populations have been diversifying in Africa for considerably longer than they have been elsewhere.

One other complicating factor in all this is that there appears to be no correlation whatever between the achievement in the human lineage of behavioral modernity and anatomical modernity. Thus, as I've just mentioned, we have evidence of humans who looked exactly like us in the Levant at close to 100 kyr ago. But at the same time, in dramatic contrast to what happened in Europe, the Levantine Neanderthals persisted in the area for some 60 kyr after the moderns appeared. What's more, throughout this long period of coexistence (whatever form it took, and frankly we have no idea how the different hominids contrived to share the landscape for all those millennia), as far as we can tell from the toolkits they made and the sites they left behind, the two kinds of hominid behaved in more or less identical ways. Suggestively, it was not until right around the time that Cro-Magnon-equivalent stoneworking techniques showed up in the Levant, at about 45 kyr ago, that the Neanderthals finally yielded possession of the area. And it was almost certainly the adoption of symbolic cognitive processes that gave our kind the final—and, for the Neanderthals, fatal—edge. The conclusion thus seems ineluctible that the emergence of anatomically modern *Homo sapiens* considerably predated the arrival of behaviorally modern humans. But while this might sound rather counterintuitive (for wouldn't it be most plausible to "explain" the arrival of a new kind of behavior by that of a new kind of hominid?), it actually makes considerable sense. For where else can any behavioral innovation become established, except within a preexisting species?

The Brain and Innovation

Nobody would dispute that to understand cognitive processes in any vertebrate species, we have to look to the brain. In the case of our own family, *Homo neanderthalensis* was endowed with a brain as large as our own, albeit housed in a skull of remarkably different shape. And while we know from the very different archaeological records they left behind that Neanderthals and Cro-Magnons behaved in highly distinctive ways, specialists on human brain evolution are hard put to identify any features on the external surface of the brain (as revealed in casts of the interior of the braincase) that would by themselves suggest any major functional difference between Neanderthal and modern *sapiens* brains. The same is obviously true for the brains of those early *sapiens* whose material cultures and ways of life resembled those of the Neanderthals. Clearly, then, we cannot attribute the advent of modern cognitive capacities simply to the culmination of a slow trend in brain improvement over time. Something happened other than a final physical buffing-up of the cognitive mechanism. Of course, by the time modern-looking humans came on the scene the necessary groundwork must have been laid for the adoption of modern cognitive processes; but this is not necessarily the same as saying that a specific neural mechanism had been acquired for them.

Let's look again, for a moment, at what our knowledge of the evolutionary process suggests may have occurred. First, it's important to remember that new structures do not arise *for* anything. They simply come about spontaneously, as byproducts of copying errors that routinely occur as genetic information is passed from one generation to the next. Natural selection is most certainly not a generative force that calls new structures into existence; it can only work on variations that are presented to it, whether to eliminate unfavorable variants or to promote successful ones. We like to speak in terms of "adaptations," since this helps us to make up stories about how and why particular innovations have arisen, or have been successful, in the course of evolution; but in reality, all new genetic variants must come into being as exaptations. As I've noted, the difference is that, while adaptations are features that fulfill specific identifiable functions (which they cannot do, of course, until they are in place), exaptations are simply features that have arisen and are potentially available to be co-opted into some new function. This is routine stuff, for many new structures stay around for no better reason than that they just don't get in the way.

This is the general context in which we are obliged to view both the evolution of the human brain as we are familiar with it today, and the appearance of modern cognitive function. There was unquestionably an increase in average hominid brain size over the past two million years, although as I've pointed out elsewhere in this book, this doesn't tell us much about the actual events of human brain evolution. But the example of the Neanderthals and, even more tellingly, of the anatomical-but-not-behavioral moderns, shows us that the arrival of the modern cognitive capacity did not simply involve adding just a bit more neural material, that last little bit of extra brain size that pushed us over the brink. Still less did it involve adding any new brain structures, for basic brain design remains remarkably uniform among all the higher primates. Instead, an exapted brain, equipped since who knows when with a neglected potential for symbolic thought, was somehow put to use.

Unfortunately, exactly what it was that exapted the brain for modern cognitive purposes remains obscure. This is largely because, while we know a lot about brain structure and about which brain components are active during the performance of particular functions, we have no idea at all about how the brain converts a mass of electrical and chemical signals into what we are individually familiar with as consciousness and thought patterns. And it is this which it will be crucial to understand if we are ever to make the leap to comprehending exactly what it is that enables us to be (and I use the term advisedly) human.

Still, it is possible to talk in general terms about the evolution of modern cognition. It has, for example, been argued that at some time between, say, 60 and 50 kyr ago, a speciation event occurred in the human lineage that gave rise to a new, symbolically expressive entity. By implication this new species would have possessed neural modifications that permitted modern behavior patterns. It would be nice to believe this, because on one level it would certainly simplify the story. The problem is, though, that the time frame doesn't appear to permit it. For this explanation to work, a new human species, physically identical but intellectually superior to one that already existed, would have had to appear and then to spread throughout the Old World in a remarkably short space of time, totally eliminating its predecessor species in the process. And there is no indication at all, in an admittedly imperfect record, that anything of this kind occurred. Which leaves us with only one evident alternative.

Instead of some anatomical innovation, perhaps we should be seeking some kind of cultural stimulus to our extraordinary cognition. If the modern human brain with all its potential capacities had been born along with modern human skull structure at some time around 100–150 kyr ago (give or take), it could have persisted for a substantial amount of time as exaptation, even as the neural mass continued to perform in the old ways. As I have already lamented, we have much less evidence than we would like that directly bears on the origin and spread of *Homo sapiens*. However, we do know that our species originated in this general time frame, probably in Africa. And we know as well that it quite rapidly spread Old World–wide from its center of origin, wherever that was. Further, if at some point, say around 60–70 kyr ago, a cultural innovation occurred in one human population or another that activated a potential for symbolic cognitive processes that had resided in the human brain all along, we can readily explain the rapid spread of symbolic behaviors by a simple mechanism of cultural diffusion. It is much more convincing (and certainly more pleasant) to claim that the new form of behavioral expression spread rapidly among populations that already possessed the potential to absorb it than it is to contemplate the alternative that the worldwide distribution of the unique human capacity came about through a process of wholesale population replacement. What carnage this latter would undoubtedly have involved! On the other hand, cultural interchange among human populations is a phenomenon that is widely documented throughout recorded history, and it must clearly be the preferred explanation for the rapid success of symbolically mediated human behaviors. It remains, though, to suggest what the new cultural stimulus might have been.

Cognition and Symbolism

When we speak of "symbolic processes" in the brain or in the mind, we are referring to our ability to abstract elements of our experience and to represent them with discrete mental symbols. Other species certainly possess consciousness in some sense; but as far as we know, they live in the world simply as it presents itself to them.

Presumably, for them the environment seems very much like a continuum, rather than a place, like ours, that is divided into the huge number of separate elements to which we humans give individual names. By separating out its elements in this way, human beings are able constantly to re-create the world, and individual aspects of it, in their minds. And what makes this possible is the ability to form and to manipulate mental symbols that correspond to elements we perceive in the world within and beyond ourselves. Members of other species often display high levels of intuitive reasoning, reacting to stimuli from the environment in quite complex ways; but only human beings are able arbitrarily to combine and recombine mental symbols and to ask themselves questions such as "What if?" And it is the ability to do this, above everything else, that forms the foundation of our vaunted creativity.

Of course, intuitive reasoning still remains a fundamental component of our mental processes; what we have done is to add the capacity for symbolic manipulation to this basic ability. An intuitive appreciation of the relationships among objects and ideas is, for example, almost certainly as large a force in basic scientific creativity as is symbolic representation; and in the end it is the unique combination of the two that makes science—or art, or technology—possible. Certainly, intuitive reasoning can take you a long way just by itself, as I think it's justifiable to claim the example of the Neanderthals shows. The Neanderthals left behind precious few hints of symbolic abilities in the abundant record they bequeathed us of their lives, and it is clear that symbols were not generally an important factor in their existences. Still, their achievements were hardly less remarkable for that, and as far as we can tell *Homo neanderthalensis* possessed a mastery of the natural world that had been unexceeded in all of earlier human history. Indeed, it seems fair to regard the Neanderthals as exponents of the most complex—and in many ways admirable—lifestyle that it has ever proved possible to achieve with intuitive processes alone.

This inevitably brings up the question about the Neanderthals that everyone wants answered: Could they talk? Many people, especially looking at the spectacularly beautiful stone tools that the Neanderthals made with such skill, find it hard to believe that they couldn't. How, other than through the use of language, could such remarkable skills have been passed down over the generations? Well, not long ago a group of Japanese researchers made a preliminary stab at addressing this problem. They divided a group of undergraduates in two, and taught one half how to make a typical Neanderthal stone tool by using elaborate verbal explanations along with practical demonstrations. The other half they taught by silent example alone. One thing this experiment dramatically revealed was just how tough it is to make stone tools; some of the undergraduates never became proficient. But more remarkable still was that the two groups showed essentially no difference either in the speed at which they acquired toolmaking skills, or in the efficiency with which they did so. Apparently learning by silent example is just fine for passing along even sophisticated stone toolmaking techniques. Although this experiment involved modern humans, not Neanderthals, it does show quite forcefully that, once again, we are making a fundamental mistake by assuming that our way is the only way of doing business in the world. None of this is to suggest, of course, that the Neanderthals did not have some form of vocal communication, even quite sophisticated vocal communication. After all, such communication is common among all vertebrates. And there can be little doubt that Neanderthals spoke, in some general sense. What they almost certainly did not possess, however, is language as we are familiar with it.

Perhaps the best way of explaining this distinction is to look at vocal communication as it is practiced among our closest living relatives, the great apes. Chimpanzees have been particularly well studied in this regard, both in the field and the laboratory. Wild-living chimps have quite a wide repertoire of calls (at least thirty have been identified, and some are used in combination), but all of them seem to be quite closely related to emotional states; they involve no elements of "explanation," but instead express simply how the individual is feeling at the moment. Indeed, as Jane Goodall has put it, chimpanzee vocalizations are so closely tied to immediate emotional states that "the production of a sound in the *absence* of the appropriate emotional state seems to be an almost impossible task." Human language, on the other hand, while often employed to express emotion, is independent of such states, and relies on complex rules of grammar and syntax that govern the meanings conveyed. It is a product of the "higher" centers of the brain, rather than of the lower, "limbic" structures that mediate the vocal productions of wild chimpanzees. But, it has been asked, might chimpanzee vocalizations be limited not by brain functions as such but by a lack of the necessary peripheral vocal equipment possessed by humans?

To answer this question, several investigators over the last couple of decades have used ingenious techniques to teach chimpanzees "language." Some of these have involved sign languages developed for deaf humans, others the arrangement on prepared surfaces of colored plastic chips representing words, and others have involved quite sophisticated computer interfaces. Yet, whatever the medium, the chimpanzees have shown consistent limitations in communicating in this way, even after the most intensive training. Some have been able to compile quite extensive vocabularies of different "words"; but the way in which they arrange these elements is totally different from the way in which humans proceed when communicating. Even where apes are able to string together several words ("Patty want apple," for example) to make a coherent request, they are handicapped by the essentially additive nature of their utterances. It's fine to put two words together if "give ball" is all that needs to be conveyed; but more complex statements are severely limited by the number of different individual elements that need to be included in such formulations.

Thus, while some linguisticians have gone so far as to dub laboratory chimpanzee utterances "protolanguage," this term is actually rather misleading. Human language is governed by its structure, which admits of endless possibilities; ape vocalization is governed by its content, which is inherently limited by its mode of expression. Chimpanzee sounds may well communicate a limited range of things that the individuals concerned want others to understand; but language, as a way of categorizing and explaining the world, and modulating thought, cuts a lot deeper than that. Indeed, it is not at all clear that vocal communication of the chimpanzee kind is even a necessary precursor to what we know as language. We may in fact be profoundly misled by the fact that, because we are alone in the world today, ape "language" is the only living indicator we have of what capacities might have been available for our fairly remote ancestors to exploit. For, even given our remarkable ability to imagine new things within contexts with which we are familiar, we human beings run into major difficulty in conceiving states of mind that lie beyond our own individual experience. It is hard enough—read, impossible—for us to understand what is going on in the head of a chimpanzee (after all, it can't tell us). How much more difficult, then, is it to reconstruct what might have been going on in the head of a Neanderthal, or a *Homo erectus,* known only from dead bones and stone tools? How did these precursors, so near to us and yet so far, perceive and experience the world from day to day? It is intellectually frustrating to have to admit that, however many fossils we find and living sites we explore, we will never know for sure. We know that our predecessors had certain features that we still share with them today, just as in other features they differed from us. But what this mix of attributes added up to on the cognitive front is entirely guesswork.

Language and the Emergence of Human Cognition

If there is one single aspect of human mental function that is more closely tied up with symbolic processes than any other, it is surely our use of language. Language is, indeed, the ultimate symbolic mental function; and it is virtually impossible to conceive of thought as we know it in its absence. For words, it is fair to say, function as the units of human thought, at least as we are aware of it. They are certainly the medium by which we explain our thoughts to one another and, as incomparably social creatures, seek to influence what is going on in one another's brains. Thus, if we are seeking a single cultural releasing factor that opened the way to symbolic cognition, the invention of language is the most obvious candidate. Indeed, it is perhaps the only plausible one that has so far proved possible to identify. What might have happened? Here we have to return to notions of exaptation, for language is a unique aptitude that doesn't seem to have emerged from protolanguage, and certainly did not do so directly. Still, it has been argued that, since the general ability to acquire language appears to be deeply and universally embedded in the human psyche, this ability must be hardwired into every healthy human brain, where it resides as a result of "normal" Darwinian processes of adaptation by natural selection.

It is certainly true that language is not reinvented in every generation, but is rather re-expressed, as every child learns his native tongue(s) as an ordinary, if astonishing, part of the process of growing up. There is, in other words, no denying the existence in the human mind of a "language instinct." What we need to explain, however, is not only how that innate instinct was acquired but also how it made such a rapid and unprecedented appearance. As we've seen, natural selection is not a creative force, and can propel nothing into existence by itself. Rather, it can only capitalize on what is already there. In a sense this makes things easier for us, since, as far as we can tell, in the emergence of symbolic thought there is no evidence of the kind of slow trend that would be expected under Darwinian selection. What must have happened, instead, is that after a long—and poorly understood—period of erratic brain expansion and reorganization in the human lineage, something occurred that set the stage for language acquisition. This innovation would have depended on the phenomenon of emergence, whereby a chance combination of preexisting elements results in something totally unexpected. The classic example of an emergent quality is water, most of whose remarkable characteristics are entirely unpredicted by those of its constituents, hydrogen and oxygen. Nonetheless, the combination of these ingredients gives rise to something entirely new, and expected only in hindsight. Together with exaptation, emergence provides a powerful mechanism in the evolutionary process, and it truly is a driving force, propelling innovation in new directions.

In the case of linguistic potential, with its innate presence among all humans today, we have to suppose that initially a neural change occurred in some population of the human lineage. This change was presumably rather minor in genetic terms, and probably had nothing whatever to do with adaptation in the classical sense. Since during early childhood development the brain rewires itself through the creation of specific pathways from undifferentiated masses of neuronal connections, it is even possible that this event was an epigenetic rather than a genetic one, dependent on developmental stimuli. Whatever the case, it certainly seems to have made no mark on the fossil record, although ultimately its impact on the archaeological traces of the Cro-Magnons and their successors was enormous. Just as the keystone of an arch is a trivial part of the structure yet is essential to the integrity of the whole, this innovation (whatever it may have been, and we are very far from understanding that) was the final physical element that needed to be in place to make possible language and symbolic thought—and all that has flowed from them, with such fateful consequences for the world. Once it was there, of course, the potential it embodied could lie fallow, simply doing no harm, until released by a cultural stimulus in one particular population. Almost certainly, though it's hard to prove, this stimulus was the invention of language. Everyone today has language, which by itself suggests that it was a highly advantageous acquisition. And if it is as advantageous as we would wish to believe, it is hardly surprising that language, and its associated symbolic behavioral patterns, were subsequently able to spread rapidly among human populations worldwide.

So much for the spread of language from its center of origin. Exactly how this fateful novelty may have been invented is a separate question, upon which it is beyond my expertise to speculate. But with the substrate for language in place, the possibilities are numerous. My favorite among them is that an initial form language may have been invented not by adults but by children. Given the fact that the brain is not a static structure like a rubber ball but is rather a dynamic entity that reorganizes itself during development (and indeed, given the right stimuli, throughout life), it is not implausible that a rudimentary precursor of language as it is familiar today initially arose in a group of children, in the context of play. Such prelanguage might have involved words—sounds—strung together with additive meaning. It is hard to imagine that, once this invention had been made, society as a whole would not have eventually adopted it. On a Japanese island macaque monkeys living along the beach were fed by researchers with sweet potatoes. These delicacies became covered with beach grit; and pretty soon, young macaques started washing them in the sea to remove the sand. It took a while for the adults to catch on: first the females, and only last the dominant males. Doubtless, some of the older and most dominant males never deigned to indulge in this behavior, preferring a familar life of grit. But a good idea is a good idea; and it is difficult to believe that, in the case of language, once the notion of associating words with objects and ideas had developed, it would not have spread quite rapidly throughout society.

Still, the transition from a nonlinguistic lifestyle to a linguistic one as we are familiar with it involved a huge cognitive and practical leap; and it seems probable that the addition of syntax may have been a separate, and later, event, though perhaps one made inevitable by the arrival of word-object associations. A single-stage progression from inarticulacy to articulate language as we know it seems more than a little implausible; and a multiple-stage process would certainly better mirror the way in which infants acquire language, with the vocabulary beginning to develop (very rapidly) first, and syntax and (later) sentence structuring following after the age of about two years. The history of the emergence of language is undoubtedly complex—indeed, this emergence only seems even possible from our perspective because we *know* it must have occurred. Subsequent to its origin, of course, language quite obviously changed, complexified, and diversified hugely, as it became ever more widely adopted among human populations. But its common structure everywhere today, independent of culture, is surely due to the fact that the underlying basis was already there in everyone, long before language itself came along.

But there still remains one other factor to be explained. To speak you need a brain that will tell your vocal tract what to do; but you also need a vocal tract that will respond appropriately to the brain's instructions. And the primitive primate vocal tract cannot respond in this way. In fact, adult human beings are the only creatures, apes included (though some birds can mimic speech), that can physically make the sounds that are essential to articulate speech. And this ability comes at a price. The principal structures that make up the vocal tract are the larynx, the structure in the neck that houses the vocal cords; the pharynx, a tube that rises above it and opens into the oral and nasal cavities; and the tongue and its associated apparatus. Basic sounds are generated at the vocal cords, and then there is further modulation of those sounds in the pharynx and allied airways above. Among typical mammals, including the apes—and newborn humans—the larynx is positioned high in the neck, and the pharynx is consequently short, limiting what can be done to modulate vocal sounds. In adult humans, in contrast, the larynx lies low in the neck, lengthening the pharynx and increasing the potential for sound modulation. The price I've mentioned is that, while the human arrangement makes a vast array of sounds possible, it also prevents simultaneous breathing and swallowing—thereby introducing the unpleasant possibility of choking to death.

This alone suggests that there must be some powerful countervailing advantage in the human conformation of the vocal tract; but the ability to speak, unfortunately, is not it. We know this because the roof of the vocal tract is also the base of the skull. Thus, where this region is preserved in fossils, we can reconstruct in general terms what the vocal tract had looked like in life. The low larynx/high pharynx combination betrays itself in a flexion of the bones of the skull base. We begin to see some evidence of such flexion in *Homo ergaster,* almost 2 myr ago; and a skull of *Homo heidelbergensis* from

Ethiopia shows that it had reached virtually its modern degree by about 600 kyr ago. A vocal tract capable of producing the sounds of articulate speech had thus been achieved among humans well over half a million years before we have any independent evidence that our forebears were using language or speaking. Clearly, then, the adult human vocal tract cannot in origin have been an adaptation "for" modern speech—though it might have conferred some advantage in the context of a "prelinguistic" form of vocal communication. So what, then, *is* it "for"? Inevitably, we have to come back to exaptation. Despite its disadvantages, basicranial flexion appeared, and it then persisted for a very long time before being capitalized upon for its linguistic qualities. Maybe over that long period it did indeed bestow certain advantages in the production of more archaic forms of speech—forms that we are hardly in a position to characterize. Or maybe it conferred some kind of benefit in terms of respiration, which is an issue that is still very poorly understood among extinct hominids. Still, whatever the case, we have to conclude that the appearance of language and its anatomical correlates was not driven by natural selection, however beneficial these innovations may appear in hindsight to have been. At present, then, there is no way we can come up with any even modestly convincing scenario of what happened in the origination of the extraordinary creature we are, without invoking the humble process of exaptation. Clearly, we are not the result of a constant and careful fine-tuning process over the millennia, and much of our history has been a matter of chance and hazard. Nature never "intended" us to occupy the position of dominance in the living world that, for whatever reasons, we find ourselves in. To a remarkable extent, we are accidental tourists as we cruise through Nature in our bizarre ways. But, of course, we are nonetheless remarkable for that. And still less are we free of responsibility.

Chapter Seven

Written in Our Genes?

One of the most obvious of our human propensities is our thirst to *explain* everything we perceive around us. We have, it seems, a deep-seated need to ask and answer the question "Why?" whether the matter at hand concerns the world outside ourselves or aspects of our own bizarre natures. Indeed, this is presumably why you are reading this book, and it's certainly one of my motivations for writing it, for writing itself is a process of discovery. What's more, even though most of us are well aware that the world is a complex, mysterious, and often unfathomable place, we nonetheless seem to be ineluctibly drawn to simple, reductionistic explanations. We are happy to see the most complex of phenomena reduced to a few rules of thumb, or better yet to only one—even when deep down we know, particularly in the case of ourselves and our motivations, that the underlying reality is likely to be considerably murkier than this implies.

Evolutionary Psychology

Nowhere is this tendency more easily observable than in the recently developed "science" of evolutionary psychology. It is hard to think of anything more resistant to all-encompassing explanations than the vagaries of human behavior; yet, even as I write, some evolutionary psychologist, somewhere, is doubtless busy reducing yet another aspect of our comportment to the action of genes originally selected for in an ancient and mythical "environment of evolutionary adaptedness." I use the term "mythical" advisedly, for this perspective seems to harken back, almost consciously, to a golden age in which human genes were harmoniously in tune with the environment. And, of course, it suggests that, as biological beings, we are thoroughly out of kilter with the changed circumstances in which we live today. But comfortable as such visions may be compared to present experience (they irresistably bring to my mind visions of a Poussin painting, or even a Watteau, although the reality of our hunting/gathering past was probably very different, with lives better described by Shakespeare: "nasty, brutish and short"), they are hugely misleading. The monolithic "ancestral environment" they evoke is little more than a figment of our nostalgia for an idealized past that never existed. Of course, much of what I've just said is hyperbole in reaction to an interpretation of our natures that I conclude is profoundly misleading; but it is certainly the case that our behaviors are not directly programmed by our genes, and that the notion of a specifiable environment of evolutionary adaptedness doesn't correspond to any identifiable reality. It also betrays a misunderstanding of the fundamental realities of the evolutionary process.

The basic idea behind evolutionary psychology is that human behaviors are at least largely determined by the genes, the basic elements of human heredity as well as that of all other living creatures. Genes are segments of the DNA that resides in the nuclei of almost all of our cells; and they are responsible not only for directing the construction of each new individual but for the transfer of hereditary information between generations. What is most important to bear in mind about them is that in most instances there is not a direct correspondence of one gene to one character. Most genes are active in the development of numerous characteristics, and those characteristics themselves are affected by the actions of more—often many more—than one gene. What's more, it is emerging that the major genes that regulate the development of each individual are remarkably conservative: In many cases, it is not these structural genes themselves that differ significantly among even remotely related and anatomically very different species; what is crucial is the sequence and timing of when these genes are switched on or off. Clearly, we have here a very complex situation. Also, all arguments directly relating behavior to genes run into the inevitable "gene shortage." Human beings probably possess substantially fewer than 100,000 genes,[1] far from enough to control individually the complex activities that go on in the brain to mediate behaviors, especially in addition to the multifarious other functions that genes must fulfill.

The notion of an intimate relationship between genes and behavior was actually born in the field of sociobiology. This specialty developed out of the study of insect societies in which behaviors are stereotyped, and do indeed seem to be regulated—though probably more indirectly than most sociobiologists are happy to admit—by innate factors. Human beings and their relatives, on the

---

[1]As this book goes to press, the genome gurus have definitively decided (for the moment) that human beings in fact possess far fewer genes than this; something like a mere 35,000 tops. This has given rise to much pontification in the newspapers, by many who should know better, about the supremacy of "nurture" over "nature" in the processes that determine how each new individual will turn out—for, logically, the fewer genes you have, the less specific the action of each one must be. In reality, though, genes, no less than whole organisms, live in environments of staggering complexity. Long ago we should have properly recognized the intricacy and interactiveness of developmental processes, and have abandoned the gross oversimplifications implicit in the nature/nurture dichotomy.

other hand, are a different kettle of fish entirely, exhibiting an almost infinite range of behaviors, and of variations in those behaviors. What's more, any individual human can potentially display new, unique behaviors of a kind entirely unspecified by hereditary factors. Despite the manifold differences among individuals and cultures, though, evolutionary psychologists have identified a whole slew of "universal" human comportments that are claimed to run across societies. Most of these universals have to do with the disparate roles of males and females in society, which in turn reflects the inevitable consequences of males' and females' different reproductive functions and risk factors. Most important, though, on closer examination most of these regularities seem to be based on economic universals that confront every individual, whatever his or her cultural origins or situation. They are not demonstrably under direct genetic control.

Let's look at this a little more closely. Evolutionary psychologists obviously believe that the human brain (or rather brains: On their reckoning males and females are different) is in some way hardwired to predispose for certain behaviors. As an explanation for the vast array of human comportments this formulation is clearly incomplete; but it has elicited a vast amount of fawning adulation among the media, and it has earned the approval of many members of a public that evidently responds very readily to headlines such as "Infidelity: It May Be In Our Genes." Why is this? Why do we find these oversimplifications so attractive? I particularly like my colleague Meredith Small's answer to this question: Such notions reinforce our male-female stereotypes; and clearly it is in many ways very comforting to have one's socially inculcated preconceptions backed up by "science." Men, the evolutionary psychologists say, are programmed by their evolutionary histories to have multiple sexual relationships with many women, as a way of spreading their genes about as widely as possible. The game, as the evolutionary psychologists see it, is for men to father as many children as possible in order to give their genes the best chance in the next generation. In theory, there is little cost to such male behavior, since it is the women who bear the brunt of childbearing and child rearing.

A favorite of the evolutionary psychologists is Moulay Ismail the Bloodthirsty, an old-time ruler of Morocco who is credited with fathering 888 children, and whose bloodthirsty proclivities are also presumably approved of by evolutionary psychologists as a means of eliminating the genes of others from the population. Women, in contrast, are strictly limited in the number of offspring they are able to produce; and from an evolutionary psychological point of view it is in their—and their genes'—interests to latch on to a high-status male who will help them rear the children they do have as effectively and lavishly as possible. Hence the tendency of women to bond, just as the stereotype says they will.

This ingrained difference between the sexes is close to received wisdom in our society, as the old rhyme suggests: "Higgamus hoggamus, woman's monogamous; hoggamus higgamus, man is polygamous." Of course, everybody knows numerous individuals who do not conform to the stereotype, but this fact is only of limited interest to those whose central interest lies in the genes. The genes do not have an economic or independent existence in the way that individuals have, and neither do genes possess calculating brains in the way that individuals do. Having huge numbers of offspring may not in fact always be the best way for men to get their genes into the next generation, even if this were actually what they—or their genes—mainly wanted to do. And old Moulay, for example, almost certainly had a whole agenda of political considerations in mind in procreating so lavishly, not to mention probably being more than a little unhinged. If anything is possible, however crazy, there is probably someone out there (dare I say probably a male?) who will do it. Women, on the other hand, do face the unavoidable economic imperative of supporting those offspring they do have. And to improve their children's—and their own—chances by bonding and throwing the resources of a male into their rearing is a highly rational economic decision, which over time has become reinforced by a social milieu that expects such behavior. There really is no reason to explain male-female differences by reference to a hardwired state of the brain that emerged in a mythical ancestral environment. For males and females both, then as now, there was and is much more to life than the fostering of one's genes. Indeed, the most important imperative for almost everyone is largely the economic one of staying alive and as comfortable as possible in a constantly shifting world of remarkable complexity.

Nonetheless, evolutionary psychological notions seem to exert an enduring fascination. Partly, as Small points out, this is because they conveniently fit the stereotypes we all absorb as members of our society. And partly it is because they are underpinned by "science," or at least by a view of science that happens to correspond to most people's preconceptions. This science can be claimed to have a respectable pedigree, for like every great book that stands at the source of a new science or system of belief, Charles Darwin's 1859 opus *On the Origin of Species by Means of Natural Selection* is subject to multifarious interpretations. The evolutionary psychologists have chosen to adopt the neo-Darwinian rendition that gives primacy in the evolutionary process to natural selection. In effect, humans and other organisms are viewed not as populations that are making do as best they can in a specific habitat (or variety of habitats) at any particular point in time but as members of species that are continually en route to a better future, as selection perfects their adaptations. The problem as evolutionary psychologists see it is that evolution is a gradual process, and that it takes time—lots of it—for selection to favor improved adaptation to current exigencies. That is why they look to the "environment of evolutionary adaptedness" rather than to modern circumstances to explain the relationship between genes and behaviors. Our comportments are seen as the product of a now-vanished hunting-and-gathering lifestyle, and thus, almost by definition, as somewhat out of synch with today's environment and needs.

A precursor, or at least a corollary, to the strict selectionist outlook of the evolutionary psychologists is the belief, which we've already broached, that the main goal in the evolutionary game is for every individual to spread his or her genes as effectively as possible. Indeed, the extreme "selfish gene" variant of this viewpoint more or less eliminates individual organisms as actors in the evolutionary play, where it is the "immortal" genes, individually, that perform the starring roles, vying with each other more or less directly for dominance. Complete organisms are seen as little more than vehicles for the genes, whose existence is primary. This kind of outlook can lead to bizarre results: Recently, for example, some of the more literally minded of the evolutionary psychologists have taken to defending rape as an "adaptive" behavior, since it is one more way for males to spread their genes around. This implied excuse for such appalling behavior is right up there with the Twinkie defense!

More sophisticated evolutionary psychologists recognize that genes cannot regulate behavior directly but can only do so by influencing the development and function of our "behavior organ," the brain. Some have thus proposed that the brain itself is "modular," with modules for a whole host of mental functions that are governed by natural selection. The number of such modules can, of course, conveniently be multiplied as new behaviors come under consideration. But quite apart from the fact that it is not possible to give a neuroanatomical identity to modular entities of this kind, this outlook reflects one of the most pervasive misunderstandings that exists (and not just among evolutionary psychologists) about the evolutionary process. This erroneous formulation has been called the "transformational" viewpoint, and where it goes off the rails is in totally ignoring the different, and highly important, roles of individuals, populations, and species in evolution. Essentially, the transformational model focuses exclusively on change over time in particular characteristics: increases in brain size, changes in body robustness, or shifts in behavioral tendencies, for instance. Under this perspective, physical or behavioral characteristics are treated as if they have independent existences, and are somehow disconnected from the larger organisms of which they form part.

But convenient as this is—if our interests lie in change over time in the inner ear, or in bipedalism, or in the brain, or in any of the myriad other characteristics that make up the human organism (all of them perfectly legitimate concerns, of course)—it's clear that it cannot be the whole story, or even much of it. The history of the hominid family, or of any other group, is not simply the abstract sum of evolutionary changes in a number of body systems that can conveniently be examined and followed separately through time. Every characteristic is inextricably embedded, along with thousands of others, in individual organisms; and natural selection can only vote up or down on the individual as a whole, rather than on its separate parts. Either the whole creature (or species) succeeds reproductively, warts and all, or it fails. The upshot is that if we view human evolution as a matter of the fine-tuning of our individual components, behavioral or otherwise, we are greatly misleading ourselves. The history of our kind and its relatives is, instead, a much more dramatic one. It is a story of evolutionary experimentation, of species originating, battling for survival in their environment(s), and, if they are lucky (and a substantial component of luck will usually be involved), giving rise to descendant populations. The flip side of this is, of course, that many hominid species have become extinct. But they have survived or become extinct as functioning wholes, rather than as possessors of this or that individual trait; and in their lifetimes the species to which the survivors belong will not usually have changed a great deal. Whatever the details, though, rarely if ever will the history of a species be one of a constant upward clawing, toward ever higher pinnacles of behavioral or anatomical perfection.

An Alternative View

Still, species do differ behaviorally among themselves, and the way in which individuals develop is clearly not unrelated to their genes. If human beings have not been fine-tuned over the ages, what is it that makes us different? In an earlier essay I remarked that the history of the hominids has largely been one of long episodes of monotony punctuated from time to time by brief episodes of innovation, anatomical or behavioral. What's more, while earlier hominid relatives (with the important exceptions of the first bipeds and the first toolmakers) can in the main be said to have done pretty much what their predecessors had done, if a little better, this is certainly not true of the emergence of behaviorally modern *Homo sapiens*. Our remarkable species, with its unprecedented symbolic cognitive capacities, was a truly novel entrant upon the biological scene; and from an early stage it established an entirely new way of interacting with the rest of the world. As I've already suggested, this astonishing capacity of *Homo sapiens* is a quality that is best described as emergent. That is to say, it is the result of a chance coincidence of acquisitions in the brain—for which it is probably fruitless at this stage of our knowledge to seek the "reasons." Further, the cognitive potential of this new biological conformation was probably released, after a longish fallow period, by a behavioral innovation, most plausibly—but not necessarily—the invention of language. Just as important, there is no evidence in the archaeological record of any fine-tuning that gradually (and, by extension, almost inevitably) led toward the unique modern human condition. Our predecessors were not simply junior-league versions of ourselves, and we are mistaken if we try to interpret them in our own image. It *is* possible, of course, to trawl the earlier record for the occasional expression of something that could be called symbolic; but such cases, at present at least, are remarkable for being exceptions, not the rule. Thus, while the development of our amazing human brain was certainly the result of genetic changes at some level, it is hard to specify what those exact changes might have been. All we know for sure is that the emergence of the human capacity was a short-term event, and also that it was a quite recent occurrence.

All of this, of course, throws the basics of evolutionary psychology into doubt. Yes, without question our brains are ultimately the products of our genes, and of the interactions of those genes with the larger environment. Many personality traits, for example, seem to be fixed very early in life—babies tend to emerge into the world pretty much as the people they will be for the rest of their lives—but the essential human capacity and behavioral range is something common to all of us. The "environment of evolutionary adaptedness" is a pretty poorly defined concept in terms of time; but, if pressed, most evolutionary psychologists would probably characterize it as the period, starting a bit under two million years ago, during which the human brain increased threefold in size, from being not much larger than that of an ape, to its exalted volume today. Nobody knows exactly what occurred in this history of enlargement, least of all behaviorally; but what does seem to be clear is that the most radical cognitive transformation of human beings is something that occurred well after the achievement of modern brain size. And, given what we already know about our family's past, it is to this extraordinary event that we should direct our attention in the effort to understand our uniquenesses and their origins, rather than to conventional Darwinian natural selection and to our long and undeniably fascinating evolutionary trudge.

All of this suggests that it is to ourselves as we are today, rather than to the diverse denizens or environments of our long evolutionary past, that we should turn for explanations—or at least characterizations—of who we are. Certain things we do know about ourselves, of course. Chief among these is that we have restless, questing psyches that will never cease to demand to know who we are and where we came from, and that will always seek explanations for the extraordinary phenomenon that we represent. It is toward the satisfaction of this deeply ingrained and perennially unquenched curiosity that our paleoanthropological efforts are ultimately directed. As to our mysterious selves, however, we have simply—or not so simply—to look around at *Homo sapiens* as we are in the living world. Our species is an endlessly complex and contradictory phenomenon. Our behaviors are the result of eons of evolutionary acquisitions, producing a complex "layered" brain in which our behaviors today are influenced, if not governed, by some very ancient structures indeed. On the behavioral level, human beings—individually and corporately—can be characterized by virtually any pair of antitheses that you could care to think of. Which human characteristics you consider to be most significant often boil down to a choice among those antitheses; and such choices, of course, will never get us very far in understanding ourselves. Lately, though, I've begun to believe that if there is one single feature of human beings that accounts for—or at least correlates nicely with—their often bizarre behaviors, it is a simple intolerance of boredom. Think about it: It could account for a lot.

Chapter Eight

Where Now?

Any look at the human past almost begs to be extrapolated to possible future developments, and indeed the ever-booming business of futurology has routinely relied on projections of this kind. Futurologists have traditionally tended, in fact, to look backward at least as much as much as they have looked forward. Extrapolating familiar old trends into the future has invariably been the safe option, while visions of the truly new and unanticipated have commonly been a much harder sell. Nowhere has this tendency been more dramatically on display than in predictions of the human evolutionary future, which almost invariably involve visions of huge heads precariously perched atop frail bodies. The logic, of course, seems impeccable, at least as long as we restrict ourselves to the most cherished of the "trends" that are believed to have marked human evolutionary history. This trend is, naturally, the famous increase in human brain size with time. As I mentioned a few essays back, such increase is undeniable as long as one considers no more than the starting and endpoints involved. Two million years ago, our putative predecessors had brains that were well under half the size of ours today. Indeed, they hardly exceeded the ape range. So far, so good; two million years of evolution have seen a significant increase in the average human brain size.

Yet if we look more closely, the trend evaporates. We don't know how many species were involved in this assumed transformation, or how variable individuals of these species were in brain size, or what the ages and longevities of those species were. In other words, we have no idea at all about the precise pattern of brain enlargement in human evolution. All we know is that hominids two million years ago typically had small brains, that more recently some hominids had brains of intermediate size, and that today our brains are remarkably large. This is sketchy evidence indeed on which to base notions of an implicitly linear transformation in brain size in the course of the human past. Still less is it a firm basis for future projections. And when we look at the evidence for time-related trends in other human body systems, we find that indications of linear transformation are even less substantial. Indeed, the pattern throughout, behavioral as well as anatomical, has typically been one of monotony—non-change—that has only rarely been punctuated by substantial innovation.

This pattern of episodic change makes excellent sense, though, when we consider what we actually know, rather than what we tend to assume, about the evolutionary process itself. Received wisdom teaches us that evolution is a gradual process of fine-tuning that unfolds over the eons, as those individuals who are better "adapted" to the environment hand on their heritable advantages to their offspring, generation by generation. Indeed, as long as we think exclusively in terms of the evolution of individual traits, this process seems entirely self-evident and inevitable. Yet traits are not the whole story; they cannot be. Those characteristics that we are wont to label adaptations (i.e., those that we like to make up stories about) are packaged as parts of individual organisms that are of astonishing complexity. Every individual is an integrated whole, and natural selection (which is no more than differential reproductive success among individuals) only possesses a single vote—either the individual is reproductively successful, or he or she is not. There is no mechanism by which particular characteristics—still less, genes—can be singled out for favored or disfavored treatment, without involving hundreds of others as well. What's more, individuals belong to populations, and populations belong to species, and these larger entities also have roles to play in the evolutionary process.

The one thing that we do know beyond a doubt is that widespread and successful species routinely tend to spawn local variant populations. It is these local populations that are the engines of innovation in the evolutionary process. Unlike large, randomly interbreeding populations, in which the weight of genetic inertia is sufficiently strong to inhibit innovation, they are small enough—and thereby genetically unstable enough—to incorporate true heritable novelties. Once this has happened, the processes of speciation—which are independent of those that produce new morphologies—can intervene to create new, historically independent, entities. In their turn those new units can go out into the environment and compete with others like them. And all this, of course, takes place in the context of the larger ecologies of which populations and species form the component parts.

Environments, furthermore, tend to change rapidly, at rates with which natural selection would be hard pressed to keep up. The most dramatic examples of this are furnished by the five mass extinctions that are abundantly documented in the record of the past 440 million years. In these extinction events, it is estimated, from 70 to 96 percent of all the world's species disappeared quite abruptly, for reasons that had nothing to do with the excellence or otherwise of their adaptation to the environments in which they had formerly flourished. Indeed, we can see this same phenomenon being played out today. In our own times a "sixth extinction" is rapidly carrying away alarming numbers of species, as a result of yet another new and unanticipated phenomenon on the landscape: this time the mindlessly destructive activities of *Homo sapiens.* But this, of course, is another story.

Looking Ahead

If we free ourselves from the notion of evolution as no more than a process of fine-tuning, and recognize instead the roles that populations inevitably play in the determination of evolutionary histories, we can better understand the probabilities for biological or cognitive change that may or may not await in the human evolutionary future. Start from the fact that for true innovations to arise and to become permanently incorporated into some component of the human population, it will be necessary for that population to become fragmented. That is to say, it will have to be divided up into units tiny—hence genetically unstable—enough that any heritable innovations arising in them can become "fixed." Failing this, there is no way we know of in which true innovations might in future become incorporated into the human gene pool. Yet, under the conditions that prevail today, such fragmentation is unimaginable. The human population now stands at over six billion, and it is increasing all the time. Population densities are expanding virtually everywhere, and individual human beings are incomparably more mobile than ever before. With human numbers and densities swelling in this way, the trend is exactly opposite to what is required for any meaningful evolutionary change to establish itself. And this trend is no less powerful for being a total reversal of earlier tendencies. Anyone glancing around the sidewalks of New York City will be fully aware of the diversity (superficial as it may be) that exists in our species *Homo sapiens.* This diversity is of very recent origin, for to the best of our knowledge *Homo sapiens* has been around on Earth for only about 100,000–150,000 years. In this comparatively (and absolutely) short time, human populations must indeed have undergone a process of local differentiation, with subtly different populations emerging in different parts of the world. This is precisely what we would expect for any species that has become so successful competitively, and as a result so widespread. It is far less evident, though, that this is a process we can reasonably extrapolate into the future. We have to bear in mind that this minor differentiation of our species *Homo sapiens* took place at a time when human beings were pretty thin on the ground, and when the climatic changes of the Ice Ages were routinely disrupting populations of animals and plants of all kinds. When the ice caps expanded, not only did vegetation zones move toward the equator and downslope, but sea levels dropped, creating land bridges where there had been ocean before. When the climate warmed, the reverse happened, and landmasses that had previously been continuous were divided up by rising seas. In these unstable climatic and geographic conditions, circumstances were ideal for the fragmentation of the sparsely scattered human populations, and the evolutionary consequences—the emergence of distinctive local groups—were entirely what one would have predicted.

Today, in contrast, for reasons that have a lot more to do with the unusual aspects of human behavior and technological accomplishment than with any external influences, the human species is undergoing a period of reintegration. In clear affirmation of the fact that morphological innovation and the production of new species are not at all the same thing, individuals from populations that originated all over the globe are now reproducing together, as beneficiaries of our recent and unprecedented mobility both as individuals and as populations. As a result, the clear if slow tendency is toward the blurring of any distinctions that had previously developed in isolation.

Cultural and technological factors are obviously paramount in this process; and as long as this tendency continues, there is obviously no way in which any kind of meaningful innovation can be incorporated into the human gene pool. The conclusion must then be that as long as conditions stay more or less as they are, there is no prospect of significant morphological or cognitive change in the human evolutionary future. Much as we may at least occasionally be appalled by ourselves and our works, it is totally unrealistic for us to hope that beneficent evolutionary processes will intervene to improve the breed and save us from our own excesses. We will, in other words, have to learn to live with ourselves as we are. There is no deus ex machina!

Unless, of course, conditions change. Any factor that would serve significantly to reduce and fragment the human population would reintroduce the possibility of evolutionary innovation. And such factors are—unfortunately, for all invoke disaster scenarios—not that hard to envisage. For instance, an asteroid impact, of the kind that has convincingly been implicated in the disappearance of the dinosaurs some 65 million years ago, might do the trick. This would especially be the case if the cloud of dust kicked up by such an impact up were sufficient to block out sunlight and disrupt photosynthesis worldwide for an extended period of time. Recent studies place the possibilities of a major asteroid impact at close to zero for the next 100,000 years, but in the longer term a disaster of this kind is almost inevitable. Perhaps more worrying for our generation, though, because it might potentially happen at any time, is the prospect of a worldwide epidemic of some horrifying and untreatable disease. Ironically, the denser—and hence, by conventional measures, the more successful—the human population becomes, the more vulnerable to such an epidemic it will be. To flourish, all pathogens—be they viruses, bacteria, prions, or whatever—require a host population reservoir above a critical size, and the more densely packed the better. They also do best where the affected population is unconcerned about how such disease is transmitted, as is inevitable among other species but is also far too often the case among humans as well. It is only too easy to imagine (as many have, indeed, done) a situation where, in our crowded urban and even rural environments, some less fragile and even more easily communicated virus than HIV could sweep through the highly concentrated human population. Multiple-drug-resistant tuberculosis is already looming on the horizon, and who knows what other pathogens might be following behind?

Or maybe economic factors will presage disaster. The archaeological record reveals only too eloquently how, over and over again, excessively intensive exploitation of the environment has led to economic collapse and the consequent fall of even highly sophisticated civilizations. The resulting demographic crashes have often been close to total. Another set of possibilities, of course, involves variants of the notion that our technological prowess will outstrip our ability to deal with the results of putting it into practice. There is no more powerful law of nature than that of unintended consequences. However carefully we might think out the possible results of our actions, they are likely to give rise to difficulties we hadn't thought of—and fixing secondary problems of our own making is often more difficult than addressing those presented to us by Nature. Over the longer term, pollution and toxic waste might pass a critical value, and render increasing swathes of the world's surface uninhabitable by humans; or the nuclear devastation envisaged by numerous works of fiction over the past half-century might leave the realm of the imagination and impose itself in real human experience. One could go on, but the essential point is already obvious. It is indeed possible to imagine circumstances in which significant evolutionary change in the human population might be in a position to reassert itself. However, all such scenarios are nothing short of apocalyptic, and all imply the wholesale elimination of literally billions of human beings. Far better, then, that we should use our energies to actively seek ways to maintain the status quo, and to develop means to cope with our familiar yet at the same time bizarrely unfathomable selves.

So What's in the Cards?

Well, if currently conditions don't exist for any meaningful evolutionary change in the human population, what can we reasonably expect to happen? Can we just continue along our merry if thoughtless way, or should we still be concerned about our prospects as a species? Here we are on safer—if still fairly disturbing—ground in extrapolating current tendencies. The most striking current trend among *Homo sapiens* is, of course, our exploding population size, a factor that continues to affect the quality of life of virtually everyone. Back in the eighteenth century, Thomas Malthus famously worried about his perception that the human population would inevitably outstrip the food supply available to support it—and, in doing so, laid the groundwork for Darwin's and Wallace's theories of evolution by natural selection. Since Malthus's time, the global population has mushroomed; but thanks to our technological ingenuity the supply of food has increased right along with it. Yes, starvation does exist in some pockets of the human population. But it occurs as a result of excessively intense exploitation of resources in certain restricted localities, and, much more widely, because of plain old conflict, corruption, and bad government. And these, regrettably, are factors that demand political solutions and are immune to technofixes. Only the politicians, God help us, can do anything about them. Overall, food supplies do appear to be adequate to feed the vast and increasing population of *Homo sapiens*; and there is still, apparently, plenty of room to expand those supplies, even though in the long term it's plain that on a finite planet continuous growth in anything will ultimately come up against a wall.

This is why economists tend to be so cheerful when contemplating the recent huge increase in the numbers of human beings. In the two centuries and more since they have been in business, technology has always come up with new ways of feeding an expanding population; and there is no reason to suspect that, in the near future at least, this will materially change. Indeed, economists in general tend to welcome population growth as a positive thing, largely because they fear the demographic consequences of shrinking population size. As the reproductive rate of an aging population falls, an increasing economic burden descends upon a steadily diminishing productive sector. In other words, the support of an expanding class of the retired—the nonproductive—becomes the responsibility of fewer and fewer economically active individuals. Once past a certain point, the economic consequences of this demographic imbalance become quite painful, and something which neither politicians nor economists seem to be at all happy to contemplate, still less to do anything about. The simplest solution in the short term—and human beings seem to be constitutionally short-termist—always seems to be to grow one's way out of the problem, whether by immigration or by local increase. As for the long term, that can be left to take care of itself; for, after all, as the most renowned economist of the twentieth century, John Maynard Keynes, once famously remarked, "In the long term we are all dead."

Yet while the human population itself does indeed seem to be almost infinitely expandable, even if only at the cost of substantial misery as the growth of populations in much of the world routinely outstrips economic productivity, there are other resources that are not. Such self-evidently critical factors as water and reasonably clean air aside, the most important of these is the simple requirement of space. The relationship between crowding and social pathologies has been demonstrated over and over again, not just among human beings but among social mammals of all kinds. Despite the vast numbers of human beings now alive, we are still in the stage of our population growth where concerns remain largely aesthetic rather than life-threatening. But choices loom, even though the consequences of inaction still lie largely in the future. Do we really want more concreting-over of the countryside, scarring the environment with endless vistas of yet more tacky strip malls? Do we really want to see more sites of outstanding natural beauty despoiled by the littering mobs who come to admire them? Do we really want to contemplate ever-more-appalling airport delays? Do we really want increasing pollution of the world that surrounds us, as the effluvia of our expanding population inexorably build up? Do we really want to witness ongoing rapid depletion of nonrenewable resources? Do we really want to stand idly by as our species systematically eliminates others from the environment, with the prospect that we will ultimately share the world with little more than rats, pigeons, and cockroaches? Do we really want to lose such little elbow room as we have left? Even more seriously, do we really want to ignore the warnings and face the potentially disastrous consequences of global warming? Well, as it turns out, many might, especially when they consider the short-term costs of doing anything about these and other unpleasant future prospects. And, of course, such problems are always someone else's fault—and it is they, naturally, who should pay for doing something about those problems. Further, as long as the worst aspects of these visions stay largely in the aesthetic realm (at least for those of us in the developed world), averting our eyes might still seem to be a viable tactic. But we are already nibbling at the outer edges of what we can blithely ignore. Twenty years ago, who had ever heard of road rage? Today, we read about it daily in the newspapers. And, in the context of snowballing populations, there can be little doubt about two things: first, that road rage is a fairly simple function of the sheer numbers of drivers on the roads; and, second, that this disturbing behavior is no more than a foretaste of even more unappetizing features of human social life to come as a result of increasing population densities. Space—and its use—is a critical social factor that we will ignore at our peril, and it is here that the extrapolation of present trends imposes itself most alarmingly on our prospects as a species.

Numerous studies have demonstrated that, among human beings, the maximum size of social units within which individual behaviors can be regulated purely by peer pressure stands at about 150 individuals. Go beyond this size, and elaborate mechanisms of social control become necessary to discourage antisocial behaviors among at least some members of the group. And as group size increases, so does the elaborateness of such mechanisms, along with the discontents of those who chafe under them. When we get to the huge populations of post-industrial society, social institutions are so large and impersonal that they virtually oblige many citizens to seek their identities within subsets of the larger society. Some of these groups are entirely positive and experience-affirming. Others, though, are of the poisonous "Aryan Nations" variety that seek to discriminate "us" from "them," and to drive wedges into society. What we *can* be confident of for the future is that, as population densities increase, the fissile tendencies that always lie beneath the surface of society will become increasingly difficult to control.

This, then, is generally bad news. But it's certainly not all the news that there is. It is, of course, fair to predict that we will not be able to depend on natural selection to save us from ourselves—although, who knows, we may already possess some exapted characteristic that, once released, may help us to re-achieve equilibrium. More likely, though, human beings will simply continue to be the mysterious, capricious creatures with whom we are familiar, as far into the future as we can foresee. We have not been fine-tuned for anything, and our behaviors show it. Close to four centuries ago, Alexander Pope summarized the human condition—or rather, our lack of any consistent condition—in words that are almost too familiar to need repeating here (but who could resist?):

> *Placed on the isthmus of a middle state,*
> *A being darkly wise, and rudely great:*
> *With too much knowledge for the sceptic side,*
> *With too much weakness for the stoic's pride,*
> *He hangs between; in doubt to act, or rest;*
> *In doubt to deem himself a god, or beast;*
> *In doubt his mind or body to prefer;*
> *Born but to die, and reasoning but to err...*
> *Created half to rise, and half to fall;*
> *Great lord of all things, yet a prey to all;*
> *Sole judge of truth, in endless error hurl'd:*
> *The glory, jest, and riddle of the world!*

As Pope so clearly recognized, human beings—both as individuals and as a species—are a bundle of paradoxes, with no hope of extricating themselves from the contradictions that constantly beset their existences. But does a probable lack of future biological or hardwired cognitive change to bring our wildly varying propensities into line imply a prospect of unrelieved monotony? Hardly, and certainly not if you find *Homo sapiens* as bizarrely fascinating a creature as I do. Where our remarkable species is concerned, "more of the same" is certainly not a prognostication of boredom. Of course, in the long history of the hominid family vast periods of monotony have indeed been a general hallmark; but once *Homo sapiens* came on the scene all that changed. And even though the oft-heard claim that the human brain is "still evolving" in some unspecified biological way is almost certainly totally false, there is still no reason at all for believing that nothing new is going to happen on the intellectual front. Here is another instance in which it does seem permissible to extrapolate from the recent past of our species to its probable future.

To judge from prior performance, what will happen is that we will continue to discover new ways of doing innovative things with the brain that we already have. Ever since the appearance of *Homo sapiens,* with its extraordinary symbolic capacities, we have been finding novel ways to use this amazing mechanism that lies within our heads. Who would ever have imagined, thirty thousand years ago, when the Chauvet cave had already been decorated with some of the most remarkable animal and geometrical art ever created, that we would one day be transliterating our newfound language into written form? Who could have conceived, three thousand years ago, that in the not-too-distant future we would be using our brains to write symphonies, and to record them in myriad kinds of media? Who would have dreamed, a mere three hundred years ago, that we would employ our remarkable cognitive powers to send human beings to the moon, or to design supercomputers? Indeed, thirty years ago, with computers already on stage, who would have predicted the extraordinary social phenomenon of the Internet? None of these innovations required a new, improved brain; the potential was there from the start, simply waiting to be discovered. And this process of discovery will not end with the present. Of course, just as past human accomplishments are nothing short of breathtaking, the possibilities for the future are far beyond any one individual's ability to predict. But the one thing we can be sure of is that the remarkable organ which resides in our heads will ensure us a constant supply of innovations. The apparently deeply ingrained human propensity for neophilia, the craving for the new, will guarantee that things will always be happening to grab the imagination. Most of those things will, of course, be relatively trivial; but it's safe to predict that once in a while a truly significant novelty—whether technological, artistic, social, or literary—will come along.

It will not, of course, be necessary for the brain to improve physically for this pattern of innovation to continue—though who knows what exaptations are already in there waiting to be unleashed? So welcome to the future, folks. Biologically speaking, it's already here.